AF816882

SOTHIS PRESS

Sothis Press aims to completely reimagine the public domain book with an emphasis on universal access to captivatingly designed print editions. We believe that technology will enable every single book ever published to be perpetually accessible in beautiful print form. Sothis is beauty in ubiquity.

www.sothis.press
1 Waterhouse Square, London, ECIN 2ST

Brünnow's
Spherical Astronomy

Including the Chapters on Parallax, Refraction, Aberration, Precession, and Nutation

Robert Main

SOTHIS PRESS

ISBN 978-1-0365-0999-6

Brünnow's Spherical Astronomy
is in the public domain.

BRÜNNOW'S

SPHERICAL ASTRONOMY.

Cambridge:
PRINTED BY C. J. CLAY, M.A.
AT THE UNIVERSITY PRESS.

BRÜNNOW'S

SPHERICAL ASTRONOMY,

TRANSLATED BY

THE REV. ROBERT MAIN, M.A. F.R.S.

PRESIDENT OF THE ROYAL ASTRONOMICAL SOCIETY;
AND RADCLIFFE OBSERVER AT OXFORD.

PART I.

INCLUDING THE CHAPTERS ON

PARALLAX, REFRACTION, ABERRATION, PRECESSION, AND NUTATION.

CAMBRIDGE:
DEIGHTON, BELL, AND CO.
LONDON: BELL AND DALDY.
1860.

TRANSLATOR'S PREFACE.

THE want of a Text-book for Spherical Astronomy has been long felt in the University of Cambridge, and indeed in all the Universities of the British Isles in which the Mathematical Sciences are cultivated. The treatise of Maddy, excellent as it was for the time when it was written, has long since ceased to be of much service for the instruction of students who are desirous of knowing the methods, both theoretical and practical, used by the astronomers of the present time; while that of Woodhouse, though it deals more practically with the operations pursued in fixed Observatories, is also comparatively useless, because it describes both instruments and methods which have long been obsolete. It has been, at least in part, a consequence of the want of a good treatise on practical astronomy, that this science has been far less cultivated in the University of Cambridge than it ought to have been, and that very few indeed of the students acquire any adequate acquaintance with the methods of the present time, as taught in the works of the great German and English astronomers.

Before the publication of Brünnow's *Sphärischen Astronomie* these methods were to be sought for either in the *Philosophical Transactions*, the *Memoirs of the Royal Astronomical Society*, or the German Astronomical periodicals, of which the chief and most celebrated is the *Astronomische*

Nachrichten. But, even of the comparatively few who could consult with advantage the latter work through their familiarity with the German language, only a small number would find the subjects of their research treated in a way at all adapted for academical use, where clear enunciation of methods and concise demonstrations of leading propositions are required rather than the minute refinements introduced by the practical astronomer when treating of special processes. Brünnow's work was, therefore, received in Germany with universal satisfaction, and, even in England, it has been read much more extensively than might have been expected, when we consider that comparatively few of the students of any of our Universities are acquainted with the German language, and still fewer of those who obtain distinction in mathematical acquirements.

Indeed the excellence of the work was so well recognized, that I was, very soon after its publication, requested by some of the greatest of English astronomers to undertake a translation, and I so far responded to the invitation as to complete a considerable portion of it. Want of time, however, and the absorption of a considerable portion of my leisure in works of a more original character, prevented me from completing it, and the Manuscript would still have remained neglected, if the increasing desire for some work on astronomy which should supply pressing needs had not induced the publishers to undertake the printing of it imperfect as it is.

It is hoped that it may still, although incomplete, be of considerable service, since, with the exception of the detailed accounts, given in the latter portion of the work, of the construction and use of instruments, and of the methods employed by modern astronomers for deducing from observations the most accurate values of the constants employed in astronomy, the portion printed contains nearly the whole of the work which would be valuable to English students.

In particular I would remark, that the theories of Parallax, Refraction, Aberration, Precession, and Nutation are given in a more complete form than is to be found in any English work, and that Mr Carrington has, in his admirable Introduction to the *Red Hill Catalogue of Circumpolar Stars,* expressed his obligations to the author by referring to the theory of Precession and Nutation.

From these considerations, I trust that the imperfect work now offered may meet with a sufficiently favourable reception to repay me by its utility for the generally thankless labour of translation, and that the lecturers and examiners in our Universities may find it sufficiently well adapted to their purpose to form the basis of an improved system for the teaching of Practical Astronomy. I might then be induced to undertake the labour of translating the remainder of the work, or, at least, of such a portion of it as would be useful to the English student.

ROBERT MAIN.

RADCLIFFE OBSERVATORY, OXFORD,
October 4, 1860.

TABLE OF CONTENTS.

INTRODUCTION.

A. *Transformation of Co-ordinates. Formulæ of Spherical Trigonometry.*

B. *On Interpolation.*

SPHERICAL ASTRONOMY.

FIRST SECTION.

I. *On the different Systems of Planes and Circles on the Visible Sphere of the Heavens.*

II. *On the Transformation from one to another of the several Systems of Co-ordinates.*

III. *Particular Phenomena of the Daily Motion.*

IV. *On the Daily Motion as a Measure of Time. Sidereal Time, Solar Time, Mean Time.*

SECOND SECTION.

I. *Parallax.*

ERRATA.

Page 11, line 5 from bottom,

for $\cos\frac{1}{2}a \,.\, \cos\frac{1}{2}(B+C)$ *read* $\left[\cos\frac{1}{2}a \,.\, \cos\frac{1}{2}(B+C)\right]$

— 11, line 4 from bottom,

for $\cos\frac{1}{2}a \,.\, \sin\frac{1}{2}(B+C)$ *read* $\left[\cos\frac{1}{2}a \,.\, \sin\frac{1}{2}(B+C)\right]$

— 12, middle of page,

value of $\frac{1}{2}(b-c)$, *dele* the sign *minus* before $185^0.45'.24'',13$

— 14, line 12 from bottom, *for* equations *read* equation

— 16, line 2 from bottom of note at foot of page, *for* complement *read* reciprocal

— 17, line 24 from top, *for* (4) *read* (6)

— 17, line 28 from top, *for* this *read* the same

— 19, line 16,

for $-\frac{\sin C}{\sin^2 b}db + \frac{\cos a}{\sin b}\,.\,dc$ *read* $-\frac{\sin C}{\sin b}\,.\,db + \frac{\cos a \,.\, \sin B}{\sin b}\,.\,dc$

— 22, formula (16), *after* $\sin 3x$ *insert* — &c.

— 24, fifth line from bottom, after $\sin 2c$, *for* $-$ *read* $+$ And, in the following formula, after $\sin 2(a+b)$, *for* $+$ *read* $-$

— 25, line 11, In the expansion of y, in the last term of the equation, *for* $\frac{1}{3}$ *read* $\frac{1}{6}$, and in the expression for $\frac{d^3(fz)}{dz^3}$ *dele* $\frac{1}{6}$

— 28, lines 17 and 19, *for* $f.a$ *read* $f(a)$, and, in the line following, "we obtain," *for* $fa=a$ *read* $f(a)=a$

— 34, end of line 12, *insert* fall

— 34, line 17, *for* $f'''(a+1)$ *read* $f''(a+1)$

— 38, line 6 from bottom, *for* $\frac{3}{2}f'''$ *read* $\frac{1}{3}f'''$

— 41, after "The two terms," *for* $\frac{1}{4}af^{\text{IV}}(a)$ *read* $\frac{1}{4}nf^{\text{IV}}(a)$

— 45, line 10, opposite 12^h, *for* 1,64 *read* 2,64

— 45, last line, *for* $25'.56'',77$ *read* $25^m.56^s,77$
for $2'',51$ *read* $2^s,51$

— 46, line 2, *for* $2'.9'',77$ *read* $2^m.9^s,77$

— 55, line 25, *for* (1*a*) *read* (2)

— 61, line 7 from bottom, *for* A *read* A_0

— 70, line 14 from bottom, *for* y'' *read* $-y''$

Page 70, line 2 from bottom, *for* x *read* x'
— 72, line 2 of note, *for* (d) *read* (a)
— 78, line 18 *for* $9^h. 15^m. 5^s.$ *read* $9^h. 13^m,5$
for $19^h. 3^m. 9^s.$ *read* $19^h. 3^m,9$
— 98, line 3 from bottom, *after* "consequently we have," in the numerator of the fraction on the right-hand side of equation, *for* ρ' *read* ρ
— 100, last paragraph, line 4 from bottom,

$$for\ \cos\frac{1}{2}(\alpha'+\alpha)\ \ read\ \frac{1}{2}(\alpha'+\alpha)$$

— 102, in the Example, log sin $(\gamma - \delta)$, *dele* the *minus* sign
— 110, to the second and third expressions for $d\zeta'$, *prefix* the sign $-$
— 116, equation (n), *for* $\dfrac{a^2}{1.2\ldots\sin^4 z}$ *read* $\dfrac{a^2}{1.2\sin^4 z}$
— 116, last equation, *for* $(e^{-\beta s}-1)$ *read* $(e^{-\beta u}-1)$
— 117, equation (o), third term in bracket, *for* $2.2^2 e^{-\beta s'}$ *read* $2.2^2.e^{-2\beta s'}$
— 118, equation (A), left-hand side, dt should range with e
— 120, expression for $\int_T^\infty e^{-t^2}dt$, *insert* reference-letter (a)
— 127, line 4 from bottom, *for* main *read* mean
— 131, line 1, *for* (a) *read* (a)
— 131, line 7 from bottom, *for* B *read* b
— 136, lines 8 and 10, formulæ, prefix the sign $-$
— 138, line 4 from bottom, *for* $7^h. 53^m,7$ *read* $7^h. 53^m. 7^s$
— 139, last line, *for* $(z' - z)$ *read* $(t' - t)$
— 140, line 22, *for* ζ' *read* ζ
— 144, last line of second paragraph, *for* $87^0.9$ *read* $87^0.9'$

INTRODUCTION.

A. TRANSFORMATION OF CO-ORDINATES. FORMULÆ OF SPHERICAL TRIGONOMETRY.

1. IN Spherical Astronomy we treat of the places of the heavenly bodies on the visible sphere of the heavens, by referring them by means of spherical co-ordinates to certain great circles of the sphere, and by investigating the relations between the co-ordinates referred to different great circles. The place of a body can be also given by means of polar co-ordinates as well as by spherical co-ordinates, for instance, by means of the angle which the straight lines drawn from it to the center of the visible sphere make with certain planes and of the distance from that center, which, being the radius of the sphere, is in this case the unit of distance. These *polar co-ordinates* can lastly be with ease expressed in terms of *rectangular co-ordinates.* The whole of Spherical Astronomy is for these reasons reduced to the transformation of rectangular co-ordinates, for which, first of all, it will be necessary to investigate the ordinary expressions.

Imagine in a plane two lines crossing each other at right angles as axes of co-ordinates, and let x and y denote the abscissa and ordinate of any point referred to them, and two other similar rectangular axes with reference to which u and v are the abscissa and ordinate of the same point, the two pairs of axes having a common point of intersection and making with each other respectively an angle w, then will x and y be functions of u and v and of the angle w, so that

$$x = \phi(u, v, w),$$
$$y = \psi(u, v, w).$$

Denote the co-ordinates of another point referred to the same pairs of axes by $x \pm x'$, $y \pm y'$, and $u \pm u'$, $v \pm v'$ respectively, then will

$$x \pm x' = \phi(u \pm u',\ v \pm v',\ w),$$
$$y \pm y' = \psi(u \pm u',\ v \pm v',\ w).$$

It is however easily seen, that if axes of co-ordinates be drawn through the first point parallel to the others, then

$$x \pm x' = \phi(u, v, w) \pm \phi(u', v', w),$$
$$y \pm y' = \psi(u, v, w) \pm \psi(u', v', w).$$

From these equations it follows that x and y are linear functions of u and v, and since for $u = 0$ and $v = 0$, also $x = 0$ and $y = 0$, x and y must be of the form

$$\left.\begin{aligned} x &= \alpha u + \beta v \\ y &= \gamma u + \delta v \end{aligned}\right\} \qquad (a),$$

where α, β, γ, δ are functions of w only.

To determine the values of these functions, the following equation must be employed,

$$x^2 + y^2 = u^2 + v^2,$$

from which, by the help of equations (a), the following equations of condition are obtained,

$$(\alpha^2 + \gamma^2 - 1)\, u^2 + (\beta^2 + \delta^2 - 1)\, v^2 + 2\,(\alpha\beta + \gamma\delta)\, uv = 0,$$

an equation which can in general only be satisfied by making

$$\alpha^2 + \gamma^2 = 1 \qquad (b),$$
$$\beta^2 + \delta^2 = 1 \qquad (c),$$
$$\alpha\beta + \gamma\delta = 0 \qquad (d).$$

From the equation (b) we obtain

$$\alpha \frac{d\alpha}{dw} + \gamma \frac{d\gamma}{dw} = 0,$$

an equation which again is satisfied in general, if, at the same time, we make

$$\alpha = c \frac{d\gamma}{dw}, \quad \text{and } \gamma = -c \frac{d\alpha}{dw},$$

$$\text{or } \alpha = -c\frac{d\gamma}{dw}, \quad \text{and } \gamma = c\frac{d\alpha}{dw}.$$

From the first equation then follows

$$\frac{d\gamma}{dw} = \frac{1}{c}\sqrt{(1-\gamma^2)},$$

$$\text{or} \quad \frac{dw}{d\gamma} = \frac{c}{\sqrt{(1-\gamma^2)}},$$

whence

$$\gamma = \sin\left(\frac{w}{c} - C\right), \quad \text{and } \alpha = \cos\left(\frac{w}{c} - C\right).$$

From the other two equations we should have obtained

$$\alpha = \sin\left(\frac{w}{c} - C\right), \quad \text{and } \gamma = \cos\left(\frac{w}{c} - C\right).$$

In the same manner it follows from equation (c), that β and δ are the sine and cosine of the angle $\frac{w}{c} - C$.

Since now for $w = 0$, $x = u$ and $y = v$, C must equal 0, and thus $\alpha = \cos\frac{w}{c}$, and $\gamma = \sin\frac{w}{c}$. Again, since for $w = 90^0$, $y = u$, and $x = -v$, if we reckon the angle w from the positive side of the axis of x towards the positive side of the axis of y, it follows that $c = 1$, and hence $\beta = -\sin w$, and $\delta = \cos w$.

Thus, for the transformation of rectangular co-ordinates, we have the formulæ,

$$\left.\begin{aligned} x &= u\cos w - v\sin w \\ y &= u\sin w + v\cos w \end{aligned}\right\} \quad \ldots\ldots\ldots\ldots\ldots\ldots \quad (1),$$

$$\text{or} \quad \left.\begin{aligned} u &= x\cos w + y\sin w \\ v &= -x\sin w + y\cos w \end{aligned}\right\} \quad \ldots\ldots\ldots\ldots\ldots\ldots \quad (2).$$

These formulæ serve generally for all positive or negative values of x and y, and for all values of w from 0^0 to 360^0.

2. Let x, y, z, be the co-ordinates of a point O referred to any system of rectangular axes, and let a' be the angle which the radius vector makes with its projection on the plane of x, y; B' the

angle which this projection makes with the axis of x (that is, the angle which the plane passing through the point O and the positive axis of x makes with the plane passing through the positive axes of x and z, reckoned from the positive side of the axis of x towards the positive side of the axis of y from 0° to 360°); then, if the distance of the point from the origin of co-ordinates be taken for the unit of distance,

$$x = \cos B' \cos a', \quad y = \sin B' \cos a', \quad z = \sin a'.$$

Again, if we denote by a the angle which the radius vector makes with the positive axis of z, reckoned from the positive side of the axis of z towards the positive sides of the axes of x and of y from 0° to 360°, we have

$$x = \sin a \cos B', \; y = \sin a \sin B', \; z = \cos a.$$

Imagine now a second system of co-ordinates such that the axis of v coincides with the axis of y, and the axes of u and w make an angle c with the axes of x and z, and call b the angle which the radius vector makes with the positive axis of w, A' the angle which the plane passing through O and the positive axis of w makes with the plane which passes through the positive axes of x and z, both angles being reckoned in the same directions as a and B', we have then

$$u = \sin b \cos A, \; v = \sin b \sin A, \; w = \cos b.$$

And since, by the formulæ for the transformation of co-ordinates,

$$z = u \sin c + w \cos c,$$
$$y = v,$$
$$x = u \cos c - w \sin c,$$

we obtain

$$\left.\begin{aligned} \cos a &= \sin b \sin c \cos A' + \cos b \cos c \\ \sin a \sin B' &= \sin b \sin A' \\ \sin a \cos B' &= \sin b \cos c \cos A' - \cos b \sin c \end{aligned}\right\} \ldots\ldots (3).$$

3. Imagine now a sphere to be described about the origin of co-ordinates with any radius whatsoever (here taken as the

unit of linear measure) and the points of intersection of the axes of z and w to be connected with each other and with the point O above mentioned by arcs of great circles, then will these arcs form a spherical triangle, taking the same in its most general signification, when the angles as well as the sides may be greater than 180°. The three sides OZ, OW, and WZ of this spherical triangle are respectively equal to a, b, and c. The spherical angle A at the point W, being the angle between the plane passing through O and W and that through W and Z and the center of the sphere, will be equal to A'; and in like manner the angle B at the point Z will be equal to $180° - B'$. Put now A and B instead of A' and B' in the equations found in (3) and we obtain the following formulæ applicable to any spherical triangle,

$$\cos a = \cos b \cos c + \sin b \sin c \cos A,$$

$$\sin a \sin B = \sin b \sin A,$$

$$\sin a \cos B = \cos b \sin c - \sin b \cos c \cos A.$$

These are the three fundamental formulæ of spherical trigonometry, which thus express nothing more than a simple transformation of co-ordinates.

Since now we may look upon any angle of a spherical triangle as the projection of the point O on the surface of a sphere and the two others as the points of intersection of the axes of z and w with the same, it follows that the foregoing formulæ must serve for any other sides and the adjacent angle by simply interchanging the remaining sides and angles amongst each other. Thus we obtain, by collecting all the cases,

$$\left.\begin{aligned}\cos a &= \cos b \cos c + \sin b \sin c \cos A\\ \cos b &= \cos a \cos c + \sin a \sin c \cos B\\ \cos c &= \cos a \cos b + \sin a \sin b \cos C\end{aligned}\right\}\ \ldots\ldots\ldots\ (4),$$

$$\left.\begin{aligned}\sin a \sin B &= \sin b \sin A\\ \sin a \sin C &= \sin c \sin A\\ \sin b \sin C &= \sin c \sin B\end{aligned}\right\}\ \ldots\ldots\ldots\ldots\ldots\ (5);$$

$$\left.\begin{aligned}
\sin a \cos B &= \cos b \sin c - \sin b \cos c \cos A \\
\sin a \cos C &= \cos c \sin b - \sin c \cos b \cos A \\
\sin b \cos A &= \cos a \sin c - \sin a \cos c \cos B \\
\sin b \cos C &= \cos c \sin a - \sin c \cos a \cos B \\
\sin c \cos A &= \cos a \sin b - \sin a \cos b \cos C \\
\sin c \cos B &= \cos b \sin a - \sin b \cos a \cos C
\end{aligned}\right\} \ldots\ldots\ (6).$$

4. From these formulæ the remaining formulæ of spherical trigonometry are easily deduced. By dividing the formulæ (6) by the corresponding formulæ (5), we obtain,

$$\left.\begin{aligned}
\sin A \operatorname{cotan} B &= \operatorname{cotan} b \sin c - \cos c \cos A \\
\sin A \operatorname{cotan} C &= \operatorname{cotan} c \sin b - \cos b \cos A \\
\sin B \operatorname{cotan} A &= \operatorname{cotan} a \sin c - \cos c \cos B \\
\sin B \operatorname{cotan} C &= \operatorname{cotan} c \sin a - \cos a \cos B \\
\sin C \operatorname{cotan} A &= \operatorname{cotan} a \sin b - \cos b \cos C \\
\sin C \operatorname{cotan} B &= \operatorname{cotan} b \sin a - \cos a \cos C
\end{aligned}\right\} \ldots\ldots\ (7).$$

If we write the last of these equations thus,

$$\sin C \cos B = \frac{\cos b \sin a \sin B}{\sin b} - \cos a \sin B \cos C,$$

we obtain

$$\sin C \cos B = \cos b \sin A - \cos a \sin B \cos C,$$

or

$$\sin A \cos b = \cos B \sin C + \sin B \cos C \cos a,$$

an equation which corresponds to the first of equations (6), and contains the angles instead of the sides, and *vice versâ*. By changing the letters the following six equations are obtained:

$$\left.\begin{aligned}
\sin A \cos b &= \cos B \sin C + \sin B \cos C \cos a \\
\sin A \cos c &= \cos C \sin B + \sin C \cos B \cos a \\
\sin B \cos a &= \cos A \sin C + \sin A \cos C \cos b \\
\sin B \cos c &= \cos C \sin A + \sin C \cos A \cos b \\
\sin C \cos a &= \cos A \sin B + \sin A \cos B \cos c \\
\sin C \cos b &= \cos B \sin A + \sin B \cos A \cos c
\end{aligned}\right\} \ldots\ldots\ (8),$$

and, dividing these equations by the corresponding ones of equations (5), we obtain

$$\left.\begin{aligned}
\sin a \cot b &= \cot B \sin C + \cos C \cos a\\
\sin a \cot c &= \cot C \sin B + \cos B \cos a\\
\sin b \cot a &= \cot A \sin C + \cos C \cos b\\
\sin b \cot c &= \cot C \sin A + \cos A \cos b\\
\sin c \cot a &= \cot A \sin B + \cos B \cos c\\
\sin c \cot b &= \cot B \sin A + \cos A \cos c
\end{aligned}\right\} \quad \ldots\ldots\ (9).$$

The equations (8) give in addition

$$\cos A \sin C = \sin B \cos a - \sin A \cos C \cos b,$$
$$\cos B \sin C = \sin A \cos b - \sin B \cos C \cos a.$$

Multiplying both equations by $\sin C$ and substituting the value of $\sin A \sin C \cos b$ from the second of the equations in the first, we obtain

$$\cos A = \sin B \sin C \cos a - \cos B \cos C,$$

and, by changing the letters, the three equations corresponding to the formula (4), in which angles take the place of sides and *vice versâ*, become

$$\left.\begin{aligned}
\cos A &= \sin B \sin C \cos a - \cos B \cos C\\
\cos B &= \sin A \sin C \cos b - \cos A \cos C\\
\cos C &= \sin A \sin B \cos c - \cos A \cos B
\end{aligned}\right\} \quad \ldots\ldots\ (10).$$

5. By adding the first two of formulæ (5), we obtain

$$\sin a (\sin B + \sin C) = \sin A (\sin b + \sin c),$$

or

$$\sin \tfrac{1}{2} a \cos \frac{B-C}{2} \cos \tfrac{1}{2} a \sin \frac{B+C}{2} = \sin \tfrac{1}{2} A \cos \frac{b-c}{2} \cos \tfrac{1}{2} A \sin \frac{b+c}{2},$$

and, by subtracting the same equations,

$$\sin \tfrac{1}{2} a \sin \frac{B-C}{2} \cos \tfrac{1}{2} a \cos \frac{B+C}{2} = \sin \tfrac{1}{2} A \cos \frac{b+c}{2} \cos \tfrac{1}{2} A \sin \frac{b-c}{2}.$$

In the same manner we obtain, by adding and subtracting the first two of formulæ (6),

$$\sin \tfrac{1}{2} a \cos \frac{B-C}{2} \cos \tfrac{1}{2} a \cos \frac{B+C}{2} = \sin \tfrac{1}{2} A \sin \frac{b+c}{2} \sin \tfrac{1}{2} A \cos \frac{b+c}{2},$$

$$\sin\frac{1}{2}a\sin\frac{B-C}{2}\cos\frac{1}{2}a\sin\frac{B+C}{2}=\cos\frac{1}{2}A\sin\frac{b-c}{2}\cos\frac{1}{2}A\cos\frac{b-c}{2}.$$

These four formulæ contain the equations of Gauss, multiplied into each other two by two; we are not able to obtain separately the single equations by the combination of these four formulæ, but for that purpose another formula must be contrived in which another combination of these equations is introduced. To this end the following equation is necessary,

$$\cos\frac{1}{2}a\cos\frac{B+C}{2}\cos\frac{1}{2}a\sin\frac{B+C}{2}=\sin\frac{1}{2}A\cos\frac{b+c}{2}\cos\frac{1}{2}A\cos\frac{b-c}{2},$$

which is obtained by adding the first two of the equations (8) together.

Put now

$$\sin\frac{1}{2}A\sin\frac{b+c}{2}=\alpha,$$

$$\sin\frac{1}{2}A\cos\frac{b+c}{2}=\beta,$$

$$\cos\frac{1}{2}A\sin\frac{b-c}{2}=\gamma,$$

$$\cos\frac{1}{2}A\cos\frac{b-c}{2}=\delta;$$

and

$$\sin\frac{1}{2}a\cos\frac{B-C}{2}=\alpha',$$

$$\cos\frac{1}{2}a\cos\frac{B+C}{2}=\beta',$$

$$\sin\frac{1}{2}a\sin\frac{B-C}{2}=\gamma',$$

$$\cos\frac{1}{2}a\sin\frac{B+C}{2}=\delta',$$

and we shall then have the five equations,

$$\alpha'\delta'=\alpha\delta,\ \gamma'\beta'=\gamma\beta,\ \alpha'\beta'=\alpha\beta,\ \gamma'\delta'=\gamma\delta,\ \beta'\delta'=\beta\delta,$$

from which the following are derived,

$$\alpha'=\alpha,\ \beta'=\beta,\ \gamma'=\gamma,\ \delta'=\delta,$$

or

$$\alpha' = -\alpha,\ \beta' = -\beta,\ \gamma' = -\gamma,\ \delta' = -\delta.$$

We consequently obtain between the angles and sides of a spherical triangle the following relations,

$$\left.\begin{aligned}
\sin\frac{1}{2}A\sin\frac{b+c}{2} &= \sin\frac{1}{2}a\cos\frac{B-C}{2}\\
\sin\frac{1}{2}A\cos\frac{b+c}{2} &= \cos\frac{1}{2}a\cos\frac{B+C}{2}\\
\cos\frac{1}{2}A\sin\frac{b-c}{2} &= \sin\frac{1}{2}a\sin\frac{B-C}{2}\\
\cos\frac{1}{2}A\cos\frac{b-c}{2} &= \cos\frac{1}{2}a\sin\frac{B+C}{2}
\end{aligned}\right\}\ \ldots\ldots\ (11),$$

or also

$$\sin\frac{1}{2}A\sin\frac{b+c}{2} = -\sin\frac{1}{2}a\cos\frac{B-C}{2},$$

$$\sin\frac{1}{2}A\cos\frac{b+c}{2} = -\cos\frac{1}{2}a\cos\frac{B+C}{2},$$

$$\cos\frac{1}{2}A\sin\frac{b-c}{2} = -\sin\frac{1}{2}a\sin\frac{B-C}{2},$$

$$\cos\frac{1}{2}A\cos\frac{b-c}{2} = -\cos\frac{1}{2}a\sin\frac{B+C}{2}.$$

From the two systems of equations we obtain for the quantities sought, whether these be two sides and the included angle, or two angles and the adjacent side, either the same values or values differing by 360°. For example, if A, b, and c be required, we shall find either for $\frac{b+c}{2}$ and $\frac{b-c}{2}$ the same values from the second set of equations as from the first, for $\frac{1}{2}A$ a value differing by 180°, or, for $\frac{b+c}{2}$ and $\frac{b-c}{2}$ values differing by 180°, and on the contrary for $\frac{1}{2}A$ the same value. Thus also A and b will always differ only by 360° from the values formed from the first system of equations. The four formulæ (11) are therefore quite general, and it is indifferent, whether, in the computation of A, b, and c the values a, B, C, be employed, or the values of these quantities $\pm 360^\circ$.*

The four equations (11) are known by the name of the *Gaussian Equations*, and are employed when one side and two adjacent angles of a spherical triangle, or two sides and the included angle, are given to find the three remaining parts. They are most easily used in the following manner. If a, B, and C are given, we first find the values of

$$[1]\ \cos\frac{B-C}{2}. \qquad [4]\ \cos\frac{B+C}{2}.$$

$$[2]\ \sin\frac{1}{2}a. \qquad [5]\ \cos\frac{1}{2}a.$$

$$[3]\ \sin\frac{B-C}{2}. \qquad [6]\ \sin\frac{B+C}{2},$$

and thence

$$[7]\ \sin\frac{1}{2}a\cos\frac{B-C}{2}. \qquad [9]\ \sin\frac{1}{2}a\sin\frac{B-C}{2}.$$

$$[8]\ \cos\frac{1}{2}a\cos\frac{B+C}{2}. \qquad [10]\ \cos\frac{1}{2}a\sin\frac{B+C}{2}.$$

By division of the numbers standing under each other we obtain $\tan\frac{1}{2}(b+c)$ and $\tan\frac{1}{2}(b-c)$, whence we find b and c. We then find the values of $\cos\frac{1}{2}(b+c)$ or $\sin\frac{1}{2}(b+c)$ and $\cos\frac{1}{2}(b-c)$ or $\sin\frac{1}{2}(b-c)$, accordingly as the sine or the cosine is the greater, and subtract the first from the greater of the two logarithms [7] or [8], the other from the greater of the two logarithms [9] or [10], and then the values of $\sin\frac{1}{2}A$ and $\cos\frac{1}{2}A$ are obtained. Combining these to find $\tan\frac{1}{2}A$, A itself is found. Since $\sin\frac{1}{2}A$ and $\cos\frac{1}{2}A$ must give the same angle as $\tan\frac{1}{2}A$, there exists a proof of the accuracy of the computation.

* Gauss, *Theoria motûs Corporum Cœlestium*, p. 50 seq.

Example.

Let

$$a = 11^\circ.\,25'.\,56'',3,$$
$$B = 184^\circ.\ \ 6'.\,55'',4,$$
$$C = 11^\circ.\,18'.\,40'',3,$$

then we have

$$\frac{1}{2}(B+C) = 97^\circ.\,42'.\,47'',85,$$

$$\frac{1}{2}(B-C) = 86^\circ.\,24'.\ \ 7'',55;$$

$$\log\cos\frac{1}{2}(B-C) = 8{\cdot}7976413,$$

$$\log\cos\frac{1}{2}(B+C) = -\,9{\cdot}1278046;$$

$$\log\sin\frac{1}{2}a = 8{\cdot}9982605,$$

$$\log\cos\frac{1}{2}a = 9{\cdot}9978351;$$

$$\log\sin\frac{1}{2}(B-C) = 9{\cdot}9991432,$$

$$\log\sin\frac{1}{2}(B+C) = 9{\cdot}9960526;$$

$$\log[\sin\frac{1}{2}a\cos\frac{1}{2}(B-C)] = 7{\cdot}7959018$$

$$\log[\sin\frac{1}{2}a\sin\frac{1}{2}(B-C)] = 8{\cdot}9974037;$$

$$\log[\cos\frac{1}{2}a\cos\frac{1}{2}(B+C)] = -\,9{\cdot}1256397,$$

$$\log[\cos\frac{1}{2}a\sin\frac{1}{2}(B+C)] = 9{\cdot}9938877;$$

$$\frac{1}{2}(b+c) = 177^\circ.\,19'.\,13'',49,$$

$$\frac{1}{2}(b-c) = 5^\circ.\,45'.\,24'',13;$$

$$\log\cos\frac{1}{2}(b+c) = -\,9{\cdot}9995248,$$

$$\log\cos\frac{1}{2}(b-c) = 9{\cdot}9978042\,;$$

$$\log\sin\frac{1}{2}A = 9{\cdot}1261149, \qquad b = 183^\circ.\,4'.\,37'',62\,;$$

$$\log\cos\frac{1}{2}A = 9{\cdot}9960835, \qquad c = 171^\circ.33'.49'',36\,;$$

$$\frac{1}{2}A = 7^\circ.\,40'.\,59'',38, \quad A = 15^\circ.21'.58'',76.$$

If we had taken

$$B = -\,175^\circ.\,53'.\,4'',6\,;$$

then

$$\frac{1}{2}(B+C) = -\,82^\circ.\,17'.\,12'',15,$$

$$\frac{1}{2}(B-C) = -\,93^\circ.\,35'.\,52'',45,$$

and thus we should have obtained

$$\frac{1}{2}(b+c) = -\quad 2^\circ.40'.\,46'',51,$$

$$\frac{1}{2}(b-c) = +\,185.\,45'.\,24'',13,$$

and thus

$$b = 183^\circ.\,4'.\,37'',62, \text{ and } c = -\,188^\circ.\,26'.\,10'',64.$$

Dividing the Gaussian Equations by each other, we obtain Napier's Analogies. Writing A, B, C, in the places of B, C, A, and a, b, c in the places of b, c, a, we find from equations (11)

$$\tan\frac{A+B}{2} = \frac{\cos\frac{a-b}{2}}{\cos\frac{a+b}{2}}\cot\text{an}\,\frac{C}{2},$$

$$\tan\frac{A-B}{2} = \frac{\sin\frac{a-b}{2}}{\sin\frac{a+b}{2}}\cot\text{an}\,\frac{C}{2},$$

$$\tan\frac{a+b}{2} = \frac{\cos\frac{A-B}{2}}{\cos\frac{A+B}{2}}\tan\frac{c}{2},$$

$$\tan\frac{a-b}{2} = \frac{\sin\dfrac{A-B}{2}}{\sin\dfrac{A+B}{2}}\tan\frac{c}{2}.$$

6. Since nearly all the formulæ in Nos. 3 and 4 consist of two terms, they are unfit for logarithmetical computation; and must be transformed into expressions having only a single term by means of auxiliary angles.

Any two possible quantities, whether positive or negative, may be made proportional to a sine or cosine, so that

$$x = m\sin M, \quad \text{and} \quad y = m\cos M;$$

from which are derived

$$\tan M = \frac{x}{y}, \quad \text{and} \quad m = \sqrt{x^2 + y^2};$$

and thus M and m are expressed in terms of possible quantities only. Now all the preceding formulæ contain in each of their two terms the sine or cosine of one and the same angle. If then we make the remaining factors of one of the terms proportional to the sine and the other proportional to the cosine of an angle, we shall be enabled to apply the formulæ for the sine or cosine of a trigonometrical quantity consisting of two terms, and by this means obtain a form convenient for logarithmetical computation.

Let for example the three formulæ for computation be

$$\cos a = \cos b\cos c + \sin b\sin c\cos A,$$
$$\sin a\sin B = \sin b\sin A,$$
$$\sin a\cos B = \cos b\sin c - \sin b\cos c\cos A,$$

and make

$$\sin b\cos A = m\sin M,$$
$$\cos b = m\cos M.$$

We shall have then

$$\cos a = m\cos(c - M),$$
$$\sin a\sin B = \sin b\sin A,$$
$$\sin a\cos B = m\sin(c - M).$$

If the quadrant in which B lies be known, the formulæ may be written in the following manner, putting for m its value $\frac{\sin b \cos A}{\sin M}$. First compute

$$\tan M = \tan b \cos A,$$

and then find

$$\tan B = \frac{\tan A \sin M}{\sin (c - M)},$$

$$\tan a = \frac{\tan (c - M)}{\cos B}.$$

Another transformation may also be effected of the formulæ

$$\cos a = \cos b \cos c + \sin b \sin c \cos A \quad \ldots\ldots\ldots (a),$$
$$\sin a \sin B = \sin b \sin A \quad \ldots\ldots\ldots (b),$$
$$\sin a \cos B = \cos b \sin c - \sin b \cos c \cos A \quad \ldots\ldots\ldots (c),$$

which will also be used in the sequel*.

Denote by B_0 and b_0 the values of B and b when $a = 90°$, we have then the three following equations,

$$0 = \cos b_0 \cos c + \sin b_0 \sin c \cos A \quad \ldots\ldots\ldots (d),$$
$$\sin B_0 = \sin b_0 \sin A \quad \ldots\ldots\ldots (e),$$
$$\cos B_0 = \cos b_0 \sin c - \sin b_0 \cos c \cos A, \quad \ldots\ldots\ldots (f).$$

Multiply (f) by $\cos c$ and subtract from the product equations (d) after multiplying the latter by $\sin c$; then multiply equation (f) by $\sin c$, and add to it equation (d) multiplied by $\cos c$, and we obtain

$$\left.\begin{aligned} \cos c \cos B_0 &= -\sin b_0 \cos A \\ \sin c \cos B_0 &= \cos b_0 \\ \sin B_0 &= \sin b_0 \sin A \end{aligned}\right\} \quad \ldots\ldots\ldots (A).$$

If then we put

$$\left.\begin{aligned} \cos c &= \sin \gamma \cos G \\ \sin c \cos A &= \sin \gamma \sin G \\ \sin c \sin A &= \cos \gamma \end{aligned}\right\} \quad \ldots\ldots\ldots (B),$$

we obtain from (d)

$$0 = \sin \gamma \cos (b_0 - G),$$

* Encke, *Jahrbuch* for 1831.

also

$$b_0 = 90^\circ + G,$$

and from (a)

$$\cos a = \sin\gamma \cos(G - b).$$

If also from the product of the equations (b) and (f) the product of (e) and (c) be subtracted

$$\sin a \sin(B - B_0) = \sin c \sin A \sin(b - b_0),$$
$$= -\cos\gamma \cos(G - b),$$

and, in the same manner, if, to the product of equations (c) and (f) the product of equations (b) and (e) and of (a) and (b) be added,

$$\sin a \cos(B - B_0) = \sin b \sin b_0 \sin^2 A + \sin b \sin b_0 \cos^2 A + \cos b \cos b_0,$$
$$= \cos(b - b_0) = -\sin(G - b).$$

The complete system of formulæ for the computation of a and B is the following,

$$\sin\gamma \cos G = \cos c,$$
$$\sin\gamma \sin G = \sin c \cos A,$$
$$\cos\gamma = \sin c \sin A,$$
$$\cos B_0 \sin c = -\sin G,$$
$$\cos B_0 \cos c = -\cos G \cos A,$$
$$\sin B_0 = \cos G \sin A,$$
$$\cos a = \sin\gamma \cos(G - b),$$
$$\sin a \sin(B - B_0) = -\cos\gamma \cos(G - b),$$
$$\sin a \cos(B - B_0) = -\sin(G - b).$$

7. In general care must be taken that the angles that are required be found by means of their tangents; for, since the variations of these functions are most rapid, the value of the angle can be found by means of them with the greatest accuracy.

Let Δx denote a very small change of an angle, then we have

$$\Delta(\log\tan x) = \frac{2\,\Delta x}{\sin 2x}.$$

It is customary to express the variations of an angle in seconds; since then the tangent is referred to radius as the unit, the variations of Δx must be also expressed in parts of the radius, and therefore be divided by the number 206264·8*. The logarithms are besides understood to be natural or hyperbolic logarithms; and, to reduce them to the system of Briggs, the former must be multiplied by the modulus $0{\cdot}4342945 = M$.

Lastly, if it is desired to express Δ (log tan x) in units of the last decimal of the logarithms which are employed, for seven-figure logarithms the expression must be multiplied by 10000000. Thus we obtain

$$\Delta\,(\log\tan x) = \frac{2M}{\sin 2x}\cdot\frac{\Delta x''}{206264{\cdot}8}\times 10000000$$

$$= \frac{42{\cdot}1}{\sin 2x}\,\Delta x'',$$

or

$$\Delta x'' = \frac{\sin 2x}{42{\cdot}1}\,.\,\Delta\,(\log\tan x).$$

From this equation it may now be seen with what precision the value of an angle can be found by means of the tangent.

If we suppose that five-figure logarithms are employed, since the computation may, in extreme cases, be incorrect to two units of the last decimal, Δ (log tan x) = 200, and the error in the angle thence arising, will be

$$\Delta\, x'' = \frac{200''}{42{\cdot}1}\sin 2x = 5''\sin 2x.$$

Thus, by the use of five-figure logarithms the error cannot be greater than $5''\sin 2x$, or, since $\sin 2x$ at its maximum is equal to unity, the greatest error may amount to $5''$, and this error can only be committed when the angle is nearly equal to 45°. With the use of seven-figure logarithms the error must be of necessity

* The number 206264·8, whose logarithm is 5·3144251, is always made use of when it is required to transform quantities expressed in parts of the radius into seconds of arc. The number of seconds in the circumference of the circle is 1296000, and the circumference of the circle in terms of the radius is $2\pi = 6{\cdot}2831853$. These two numbers are to each other in the proportion of 206264·8 : 1. If it is required to transform quantities expressed in parts of the radius into arc, they must be multiplied by this number; on the contrary, to express quantities which are given in seconds, in parts of the radius, it is necessary to divide by this number. The number itself is the number of seconds in an arc equal to the radius, and its complement is the sine or the tangent of one second.

100 times smaller, and thus the value of the angle obtained by means of the tangent, can at the most be incorrect only to $0''\cdot05$.

If now an angle be given by means of its sine or cosine, then, in the formula for

$$\Delta\,(\log \sin x) \text{ or } \Delta\,(\log \cos x),$$

we should have, instead of the factor $\sin 2x$, $\tan x$ or $\text{cotan}\, x$, each of which may have all values possible up to infinity. We thus see that small errors in the logarithms of the sine or cosine of an angle may produce very large errors in the angle sought by means of these functions, and therefore the values of angles must in preference be always sought by means of their tangents.

8. Suppose that, in the formulæ for oblique-angled spherical triangles, one of the angles is equal to 90°, we then obtain the formulæ for right-angled spherical triangles.

In what follows the hypothenuse will be constantly denoted by h, the two sides by c and c', and the angles opposite to them by C and C'.

From the first of the formulæ (4), we obtain by making $A = 90°$,

$$\cos h = \cos c \cos c'.$$

Again, from the first of the formulæ (5), on the same supposition, we obtain

$$\sin h \sin C = \sin c,$$

and from the first of the formulæ (4),

$$\sin h \cos C = \cos c \sin c',$$

or, by dividing this equation by that for the value of $\cos h$,

$$\tan h \cos C = \tan c'.$$

Again, by dividing this formula by the expression for $\sin h \sin C$, we shall have

$$\text{cotan}\, C = \text{cotan}\, c \sin c',$$

or

$$\tan c = \tan C \sin c'.$$

Both formulæ might have been obtained also from the first of equations (7) and the third of equations (9), by making $A = 90°$.

From the first of equations (10) there follows:

$$0 = \cos h \sin C \sin C' - \cos C \cos C',$$

or

$$\cos h = \text{cotan } C \text{ cotan } C'.$$

Lastly, by combining the two equations,

$$\sin h \sin C' = \sin c',$$

and

$$\sin h \cos C = \cos c \sin c',$$

we obtain

$$\cos C = \sin C' \cos c.$$

We have consequently the following six formulæ for right-angled triangles:

$$\left.\begin{aligned} \cos h &= \cos c \cos c', \\ \sin c &= \sin h \sin C, \\ \tan c &= \tan h \cos C', \\ \tan c &= \tan C \sin c', \\ \cos h &= \text{cotan } C \text{ cotan } C', \\ \cos C &= \cos c \sin C', \end{aligned}\right\} \quad \text{.................} (12),$$

by means of which, if any two parts of a right-angled triangle be given, the remaining parts may be found.

9. In astronomy it is necessary in the computation of quantities to borrow certain data derived from observation. But since the absolute accuracy of none of these data can be answered for, but in every case a small error must be looked upon as possible, it is necessary to investigate in every problem, whether a small change of the observed quantities can produce no great error in the quantities to be found. To be able to judge of this in every case, it is necessary to differentiate the formulæ of spherical trigonometry, and, that all cases may be included, all the quantities must be considered as variable.

By differentiating the first of equations (4) we obtain

$$\begin{aligned} -\sin a da = {} & db\,[-\sin b \cos c + \cos b \sin c \cos A] \\ & + dc\,[-\cos b \sin c + \sin b \cos c \cos A] \\ & - \sin b \sin c \sin A dA. \end{aligned}$$

The factor of db is equal to $-\sin a \cos C$, that of dc is equal to $-\sin a \cos B$; writing in addition $-\sin a \sin c \sin B$ instead of the factor of A, we obtain the differential formula

$$da = \cos C db + \cos B dc + \sin c \sin B dA.$$

Writing the first of equations (5) logarithmically, we obtain

$$\log \sin a + \log \sin B = \log \sin b + \log \sin A;$$

and, differentiating,

$$\operatorname{cotan} a da + \operatorname{cotan} B dB = \operatorname{cotan} b db + \operatorname{cotan} A dA.$$

In place of the first of formulæ (6), if we differentiate the first of formulæ (7), which is derived from the combination of (5) and (6), we obtain

$$-\frac{\sin A}{\sin^2 B} dB + dA\,[\operatorname{cotan} B \cos A - \sin A \cos c]$$

$$= -\frac{\sin c}{\sin^2 b} db + dc\,[\operatorname{cotan} b \cos c + \cos A \sin c],$$

or $$-\frac{\sin A}{\sin^2 B} dB - \frac{\cos C}{\sin B} dA = -\frac{\sin c}{\sin^2 b} db + \frac{\cos a}{\sin b} dc.$$

By multiplying this equation by $\sin B$, we find

$$-\frac{\sin a}{\sin b} dB - \cos C dA = -\frac{\sin C}{\sin b} db + \frac{\cos a}{\sin b} dc,$$

or finally

$$\sin a dB = \sin C db - \sin B \cos a dc - \sin b \cos C dA.$$

From the first of formulæ (10) we obtain, precisely in the same way as from (4),

$$dA = -\cos c dB - \cos b dC + \sin b \sin C da.$$

We have thus the following differential equations of spherical trigonometry:

$$\left.\begin{aligned} da &= \cos C db + \cos B dc + \sin b \sin C dA \\ \operatorname{cotan} a\, da + \operatorname{cotan} B dB &= \operatorname{cotan} b db + \operatorname{cotan} A dA \\ \sin a dB &= \sin C db - \sin B \cos a dc - \sin b \cos C dA \\ dA &= -\cos c dB - \cos b dC + \sin b \sin C da \end{aligned}\right\} (13).$$

10. For small angles it may be permitted to put unity in place of the cosine, and the arc in place of the sine or the

tangent; thus, if the arc be expressed in seconds we may put 206265 a instead of sin a or tan a.

If the angles are not small enough to allow the second term of the series for the sine to be omitted, we may proceed in the following manner:

We have

$$\frac{\sin \alpha}{\alpha} = 1 - \frac{1}{6}\alpha^2 + \frac{1}{120}\alpha^4 -$$

and

$$\cos \alpha = 1 - \frac{1}{2}\alpha^2 + \frac{1}{24}\alpha^4 -$$

thus

$$\sqrt[3]{\cos \alpha} = 1 - \frac{1}{6}\alpha^2 + \&c.$$

We have therefore to the third powers inclusive

$$\frac{\sin \alpha}{\alpha} = \sqrt[3]{\cos \alpha},$$

$$\text{or} \quad \alpha = \sin \alpha \sqrt[3]{\sec \alpha};$$

a formula, which is so accurate, that for an angle of 10° not so much as an error of 1″ would arise from its employment. In this case we have

$$\log (\sin 10^0 \sqrt[3]{\sec 10^0}) = 9{\cdot}2418864,$$

and, adding thereto the logarithm 5·3144251 and taking out the corresponding number, we obtain 36000″.74, or 10°. 0′. 0″.74.

11. In spherical trigonometry very great use is made of developments in series, of which we will here derive the most important.

If we have an expression of the form

$$\tan y = \frac{a \sin x}{1 - a \cos x},$$

we can easily develope y in a series, which shall proceed according to sines of multiples of the angle x. If, namely, $\tan z = \frac{m}{n}$,

$$dz = \frac{ndm - mdn}{m^2 + n^2}.$$

Considering then in the formula for tan y both a and y to be variable, we obtain

$$\frac{dy}{da} = \frac{\sin x}{1 - 2a \cos x + a^2};$$

and, if we develop this expression, by the method of indeterminate coefficients, in a series proceeding according to the powers of a,

$$\frac{dy}{da} = \sin x + a \sin 2x + a^2 \sin 3x + \ldots *$$

Integrating this equation, and remarking that for $x = 0$, y is also $= 0$, we obtain for y the following series:

$$y = a \sin x + \frac{1}{2} a^2 \sin 2x + \frac{1}{3} a^3 \sin 3x + \ldots\ldots\ldots (14).$$

Very frequently we have two equations of the form

$$A \sin B = a \sin x,$$
$$A \cos B = 1 - a \cos x,$$

from which it is proposed to develop B and $\log A$ in a series proceeding according to the sines and cosines of multiples of x. Since here

$$\tan B = \frac{a \sin x}{1 - a \cos x},$$

we can, by formula (14), find for B a series ascending according to sines of multiples of x. To develope $\log A$ in a similar series we have in the first place

$$A = \sqrt{1 - 2a \cos x + a^2}.$$

But, by the method of indeterminate coefficients, the following series is found:

$$\frac{a \cos x - a^2}{1 - 2a \cos x + a^2} = a \cos x + a^2 \cos 2x + a^3 \cos 3x + \text{\&c.}\dagger$$

* It is easily seen that the first term is $\sin x$, and that the coefficient of a^n will be found by the equation

$$A_n = 2A_{n-1} \cos x - A_{n-2}.$$

† It is seen immediately that the coefficient of a is equal to $\cos x$, and that the coefficient of a^n will be found by the equation

$$A_n = 2A_{n-1} \cos x - A_{n-2}.$$

By multiplying this expression by $-\frac{da}{a}$ and integrating with respect to a, we shall have, since the left side is equal to

$$\frac{1}{2}\frac{d.\log(1-2a\cos x+a^2)}{da},$$

and, for $a=0$, $\log A$, also $=0$,

$$\log\sqrt{1-2a\cos x+a^2}=\log A=-[a\cos x+\frac{1}{2}a^2\cos 2x$$
$$+\frac{1}{3}a^3\cos 3x+\&c.]\ldots\ldots\ldots(15).$$

In the same manner we obtain, if we have the two equations

$$A\sin B=a\sin x,$$
$$A\cos B=1+a\cos x,$$

by putting in (14) and (15) 180^0-x in the place of x,

$$B=a\sin x-\frac{1}{2}a^2\sin 2x+\frac{1}{3}a^3\sin 3x\ldots\ldots\ldots\ldots(16),$$

$$\log\sqrt{1+2a\cos x+a^2}=\log A=a\cos x-\frac{1}{2}a^2\cos 2x$$
$$+\frac{1}{3}a^3\cos 3x-\&c.\ldots\ldots\ldots(17).$$

If the expression be of the form

$$\tan y=n\tan x,$$

it may easily be reduced to the form

$$\tan y=\frac{a\sin x}{1-a\cos x}.$$

We have in fact

$$\tan(y-x)=\frac{\tan y-\tan x}{1+\tan y\tan x}=\frac{(n-1)\tan x}{1+n\tan^2 x}$$
$$=\frac{(n-1)\sin x\cos x}{\cos^2 x+n\sin^2 x}=\frac{(n-1)\sin x\cos x}{\frac{1}{2}+\frac{1}{2}\cos 2x+\frac{n}{2}-\frac{n}{2}\cos 2x}$$
$$=\frac{(n-1)\sin 2x}{(n+1)-(n-1)\cos 2x}=\frac{\frac{n-1}{n+1}.\sin 2x}{1-\frac{n-1}{n+1}\cos 2x}.$$

Therefore if the equation $\tan y = n \tan x$ be given, we obtain

$$y = x + \frac{n-1}{n+1}\sin 2x + \frac{1}{2}\left(\frac{n-1}{n+1}\right)^2 \sin 4x + \frac{1}{3}\left(\frac{n-1}{n+1}\right)^3 \sin 6x + \ldots (18).$$

In this expression put, in the first place,

$$n = \cos\alpha,$$

then

$$\frac{n-1}{n+1} = -\tan^2\frac{1}{2}\alpha.$$

Thus the equation

$$\tan y = \cos\alpha \tan x$$

gives

$$y = x - \tan^2\frac{1}{2}\alpha \sin 2x + \frac{1}{2}\tan^4\frac{1}{2}\alpha \sin 4x - \frac{1}{3}\tan^6\frac{1}{2}\alpha \sin 6x + \&c. \ldots\ldots (19).$$

If

$$n = \sec\alpha,$$

then

$$\frac{n-1}{n+1} = \tan^2\frac{1}{2}\alpha,$$

and we thus obtain, if

$$\tan y = \sec\alpha \tan x, \text{ or } \tan x = \cos\alpha \tan y,$$

$$y = x + \tan^2\frac{1}{2}\alpha \sin 2x + \frac{1}{2}\tan^4\frac{1}{2}\alpha \sin 4x + \frac{1}{3}\tan^6\frac{1}{2}\alpha \sin 6x + \&c. \ldots\ldots (20).$$

Since

$$\frac{\cos\alpha - \cos\beta}{\cos\alpha + \cos\beta} = \tan\frac{1}{2}(\beta-\alpha)\tan\frac{1}{2}(\beta+\alpha),$$

and

$$\frac{\sin\alpha - \sin\beta}{\cos\alpha + \cos\beta} = \tan\frac{1}{2}(\alpha-\beta)\cot\frac{1}{2}(\alpha+\beta),$$

we obtain, if

$$\tan y = \frac{\cos\alpha}{\cos\beta}\tan x,$$

$$y = x - \tan\frac{1}{2}(\alpha-\beta)\tan\frac{1}{2}(\alpha+\beta)\sin 2x$$

$$+ \frac{1}{2}\tan^2\frac{1}{2}(\alpha-\beta)\tan^2\frac{1}{2}(\alpha+\beta)\sin 4x + \ldots$$

and if
$$\tan y = \frac{\sin \alpha}{\sin \beta} \tan x,$$

$$y = x + \tan \frac{1}{2}(\alpha - \beta) \cot \frac{1}{2}(\alpha + \beta) \sin 2x$$

$$+ \frac{1}{2} \tan^2 \frac{1}{2}(\alpha - \beta) \cot^2 \frac{1}{2}(\alpha + \beta) \sin 4x + \ldots$$

By means of the last two formulæ Napier's Analogies may be developed into series. From the equation

$$\tan \frac{a-b}{2} = \frac{\sin \frac{A-B}{2}}{\sin \frac{A+B}{2}} \tan \frac{c}{2},$$

namely, we obtain

$$\frac{a-b}{2} = \frac{c}{2} - \tan \frac{B}{2} \cot \frac{A}{2} \sin c + \frac{1}{2} \tan^2 \frac{B}{2} \cot^2 \frac{A}{2} \sin 2c - \ldots$$

or

$$\frac{c}{2} = \frac{a-b}{2} + \tan \frac{B}{2} \cot \frac{A}{2} \sin (a-b) + \frac{1}{2} \tan^2 \frac{B}{2} \cot^2 \frac{A}{2} \sin 2(a-b) + \ldots$$

and in the same manner from the equation

$$\tan \frac{a+b}{2} = \frac{\cos \frac{A-B}{2}}{\cos \frac{A+B}{2}} \tan \frac{c}{2},$$

we obtain the following series,

$$\frac{a+b}{2} = \frac{c}{2} + \tan \frac{A}{2} \tan \frac{B}{2} \sin c + \frac{1}{2} \tan^2 \frac{A}{2} \tan^2 \frac{B}{2} \sin 2c + \ldots$$

$$\frac{c}{2} = \frac{a+b}{2} - \tan \frac{A}{2} \tan \frac{B}{2} \sin (a+b) + \frac{1}{2} \tan^2 \frac{A}{2} \tan^2 \frac{B}{2} \sin 2(a+b) + \ldots$$

Precisely similar series are obtained from the two other analogies

$$\tan \frac{A-B}{2} = \frac{\sin \frac{a-b}{2}}{\sin \frac{a+b}{2}} \tan \frac{180^\circ - C}{2},$$

$$\tan\frac{A+B}{2}=\frac{\cos\frac{a-b}{2}}{\cos\frac{a+b}{2}}\tan\frac{180^0-C}{2}.$$

A case frequently occurs in which a quantity y is given by means of an equation of the form

$$\cos y=\cos x+b,$$

and it is necessary to change it into a series proceeding according to powers of b. For this purpose we develop the equation

$$y=\cos^{-1}(\cos x+b),$$

by Taylor's Theorem. Make, namely,

$$\cos x=z \text{ and } y=f(z+b),$$

we have then

$$y=f(z)+\frac{d(fz)}{dz}.b+\frac{1}{2}\frac{d^2(fz)}{dz^2}.b^2+\frac{1}{3}\frac{d^3(fz)}{dz^3}b^3+\&c.$$

or since

$$f(z)=x,\ \frac{d(fz)}{dz}=\frac{dx}{d.(\cos x)}=-\frac{1}{\sin x},$$

$$\frac{d^2(fz)}{dz^2}=\frac{d.\left(-\frac{1}{\sin x}\right)}{dx}.\frac{dx}{d(\cos x)}=-\frac{\cos x}{\sin^3 x},$$

$$\frac{d^3(fz)}{dz^3}=\frac{d.\left(-\frac{\cos x}{\sin^3 x}\right)}{dx}.\frac{dx}{d(\cos x)}=-\frac{1}{6}\frac{1+3\cot^2 x}{\sin^3 x},$$

Thus

$$y=x-\frac{b}{\sin x}-\frac{1}{2}\cot x\frac{b^2}{\sin^2 x}-\frac{1}{6}(1+3\cot^2 x)\frac{b^3}{\sin^3 x}-\ldots\ldots(21).$$

Just in the same way, from the equation

$$\sin y=\sin x+b,$$

we obtain

$$y=x+\frac{b}{\cos x}+\frac{1}{2}\tan x\frac{b^2}{\cos^2 x}+\frac{1}{6}(1+3\tan^2 x)\frac{b^3}{\cos^3 x}+\ldots\ldots(22).$$

NOTE. On development of Series compare Encke, *Einige Reihenentwickelungen aus der sphärischen Astronomie. Astronomische Nachrichten.* No. 562.

B. ON INTERPOLATION.

12. In astronomy use is constantly made of tables in which the numerical values of certain functions are given for single numerical values of the variables. Since now in practice there is also need of values of the function for such values of the variables as are not given explicitly in the tables, means must be found for computing the same for any values whatever of the variable or of the argument of the function. For this purpose is employed the process of interpolation. The object of it is, to substitute in place of a function whose analytical expression is either altogether unknown or inconvenient for numerical computation, another simpler function formed from given numerical values, which, within the limits of its use, can be employed instead of the other.

By Taylor's Theorem any function can be expanded into a series proceeding according to integral powers of the variable: the only exception to this law being in the case when, for a determinate value of the variable, one of the differential coefficients becomes infinite but the function in the neighbourhood of this value is discontinuous. Since however the process of interpolation is founded on the expansion of functions in series which proceed according to integral powers of the variable, it is thus assumed that the function, within the limits in question, is continuous, and is to be employed only on this supposition.

Let w be the interval or the difference of two consecutive arguments (which here is always considered to be constant), then any argument whatever may be denoted by $a + nw$, where n is the variable quantity, and the function, corresponding to this argument, is $f(a + nw)$. The difference of two consecutive values of the function $f(a + nw)$ and $f\{a + (n + 1) w\}$ may be denoted by $f'\left(a + n + \frac{1}{2}\right)$, where, for the purpose of denoting to what values of the function the difference corresponds, the arithmetical mean of the two arguments is put under the expression for the functions and the factor w is omitted*. In the same

* This very convenient notation is given by Encke in his Treatise *On Mechanical Quadratures*, in the *Jahrbuch* for 1837.

manner $f'\left(a+\frac{1}{2}\right)$ expresses the difference of $f(a)$ and $f(a+w)$, $f'\left(a+\frac{3}{2}\right)$ the difference of $f(a+w)$ and $f(a+2w)$. The same holds also with respect to higher differences whose order is denoted by the accents. Thus, for example, $f''(a+2)$ is the difference of the first two differences $f'\left(a+\frac{1}{2}\right)$ and $f'\left(a+\frac{3}{2}\right)$.

The scheme of the arguments and of the corresponding values of the function and their differences is therefore as follows:

Arg.	Function.	1st Diff.	2nd Diff.	3rd Diff.	4th Diff.	5th Diff.
$a-3w$	$f(a-3w)$					
		$f'(a-\frac{5}{2})$				
$a-2w$	$f(a-2w)$		$f''(a-2)$			
		$f'(a-\frac{3}{2})$		$f'''(a-\frac{3}{2})$		
$a-w$	$f(a-w)$		$f''(a-1)$		$f^{\text{IV}}(a-1)$	
		$f'(a-\frac{1}{2})$		$f'''(a-\frac{1}{2})$		$f^{\text{V}}(a-\frac{1}{2})$.
a	$f(a)$		$f''(a)$		$f^{\text{IV}}(a)$	
		$f'(a+\frac{1}{2})$		$f'''(a+\frac{1}{2})$		$f^{\text{V}}(a+\frac{1}{2})$.
$a+w$	$f(a+w)$		$f''(a+1)$		$f^{\text{IV}}(a+1)$	
		$f'(a+\frac{3}{2})$		$f'''(a+\frac{3}{2})$		
$a+2w$	$f(a+2w)$		$f''(a+2)$			
		$f'(a+\frac{5}{2})$				
$a+3w$	$f(a+3w)$					

All the differences which have the same quantities under the characters of the functions, stand here in the same horizontal line. The differences of odd orders have all of them as the quantities under the function-characters, $a+$ a fraction with the denominator 2.

13. Since, by Taylor's Theorem, any function can be expanded in a series ascending by integral powers of the variable, we may make

$$f(a+nw)=\alpha+\beta\,.\,nw+\gamma\,.\,n^2w^2+\delta\,.\,n^3w^3+\&c.\;\ldots\ldots(\alpha).$$

If the analytical expression of the function $f(a+nw)$ were known, the quantities α, β, γ, δ, might be computed, since

$$\alpha=f(a),\;\; \beta=\frac{d\,.f(a)}{da},\;\; \&c.$$

It is however assumed that this analytical expression is not given, or at least, if even it is known, it is not to be employed, and that the function $f(a+nw)$ is known only for certain values of the argument $a+nw$. But if, in the above equation, we put one after another the different values of the variable n, we obtain as many equations as we know values of the function, and we can calculate just the same number of the coefficients α, β, γ, δ, &c.

Let now four numerical values of the function $f(a+nw)$ be given, namely, $f(a)$, $f(a+w)$, $f(a+2w)$ and $f(a+3w)$, we have then the four equations:

$$f(a)=\alpha,$$
$$f(a+w)=\alpha+\beta w+\gamma w^2+\delta w^3,$$
$$f(a+2w)=\alpha+2\beta w+4\gamma w^2+8\delta w^3,$$
$$f(a+3w)=\alpha+3\beta w+9\gamma w^2+27\delta w^3.$$

But since

$$f(a+w)=f(a)+f'\left(a+\frac{1}{2}\right),$$
$$f(a+2w)=f(a)+f'\left(a+\frac{1}{2}\right)+f'\left(a+\frac{3}{2}\right),$$
$$=f(a)+2f'\left(a+\frac{1}{2}\right)+f''(a+1)$$
$$f(a+3w)=f(a+2w)+f'\left(a+\frac{5}{2}\right),$$
$$=f(a)+3f'\left(a+\frac{1}{2}\right)+3f''(a+1)+f'''\left(a+\frac{3}{2}\right);$$

we obtain

$$f(a)=\alpha,$$
$$f'\left(a+\frac{1}{2}\right)=\beta w+\gamma w^2+\delta w^3,$$
$$2f'\left(a+\frac{1}{2}\right)+f''(a+1)=2\beta w+4\gamma w^2+8\delta w^3,$$
$$3f'\left(a+\frac{1}{2}\right)+3f''(a+1)+f'''\left(a+\frac{3}{2}\right)=3\beta w+9\gamma w^2+27\delta w^3,$$

and hence

$$\frac{1}{6}f'''\left(a+\frac{3}{2}\right)=\delta w^3,$$
$$\frac{1}{2}\left[f''(a)-f'''\left(a+\frac{3}{2}\right)\right]=\gamma w^2,$$
$$f'\left(a+\frac{1}{2}\right)-\frac{1}{2}f''(a)+\frac{1}{3}f'''\left(a+\frac{3}{2}\right)=\beta w,$$

and therefore, if we substitute these values in equation (a) for $f(a+nw)$ and arrange the same in the order of the differences,

$$f(a+nw) = fa + nf'\left(a+\frac{1}{2}\right) + \frac{n^2-n}{2} f''(a+1),$$
$$+ \frac{n^3-3n^2+2n}{6} f'''\left(a+\frac{3}{2}\right),$$

or

$$f(a+nw) = fa + nf'\left(a+\frac{1}{2}\right) + \frac{n.(n-1)}{1.2} f''(a+1)$$
$$+ \frac{n.(n-1)(n-2)}{1.2.3} f'''\left(a+\frac{3}{2}\right) + \ldots\ldots(1).$$

This formula is known under the name of Newton's Interpolation formula. The coefficient of the difference of the order n is the coefficient of x^n in the expansion of $(1+x)^n$. The proof, which is here given for only four values, can be easily extended to any number of values.

Example.

In the Berlin *Jahrbuch* for 1850, there are, for mean noon, the following Heliocentric longitudes of Mercury:

		1st Diff.	2nd Diff.	3rd Diff.	4th Diff.
Jan. 0	303°.25'. 1'',5				
		+6°.41'.50'',0			
2	310 . 6 .51 ,5		+18'.48'',0		
		+7. 0.38 ,0		+2'.44'',4	
4	317 . 7 .29 ,5		+21.32 ,4		+10'',1
		+7.22.10 ,4		+2 .54 ,5	
6	324 . 29.39 ,9		+24.26 ,9		+ 4 ,7
		+7.46.37 ,3		+2 .59 ,2	
8	332 . 16.17 ,2		+27.26 ,1		
		+8.14. 3 ,4			
10	340 . 30.20 ,6				

If from these the longitude be required for mean noon of January 1, we have

$$f(a) = 303^\circ.25'.1'',5, \text{ and } n = \frac{1}{2}.$$

In addition

$$f'\left(a+\frac{1}{2}\right) = +6^\circ.41'.50'',0,\ n = \frac{1}{2}.\quad \text{Product} = +3^\circ.20'.55'',0,$$

$f''(a+1) = +18'.48'',0, \dfrac{n(n-1)}{1.2} = -\dfrac{1}{8}$. Product $= -2'.21'',0$.

$f'''\left(a+\dfrac{3}{2}\right) = +2.44,4, \dfrac{n.(n-1)(n-2)}{1.2.3} = +\dfrac{1}{16}$.

Product $= +10,3$.

$f^{\text{IV}}(a+2) = +10,1, \dfrac{n.(n-1)(n-2)(n-3)}{1.2.3.4} = -\dfrac{5}{128}$.

Product $= -0,4$.

There is then to be added to $f(a)$, the quantity

$$+3^0.\ 18'.\ 43'',9,$$

and we obtain for the longitude of Mercury, on Jan. 1,0,

$$306^0.\ 43'.\ 45'',4.$$

The Newtonian formula may be still more conveniently written in the following manner, which has this advantage, that the multipliers are all small fractions:

(1a).
$$f(a+nw) = f(a) + n\left[f'\left(a+\frac{1}{2}\right) + \frac{n-1}{2}\left[f''(a+1)\right.\right.$$
$$\left.\left. + \frac{n-2}{3}\left[f'''\left(a+\frac{3}{2}\right) + \frac{n-3}{4}\left[f^{\text{IV}}(a+2) + \&c.\right.\right.\right.\right.$$

If $n = \dfrac{1}{2}$, $\dfrac{n-3}{4} = -\dfrac{5}{8}$ and $\dfrac{n-3}{4}.f^{\text{IV}}(a+2) = -6'',4$.

This, added to $f'''\left(a+\dfrac{3}{2}\right)$ and the sum multiplied by $\dfrac{n-2}{3}\left(=-\dfrac{1}{2}\right)$, gives $-1'.19'',0$. Adding this again to $f''(a+1)$ and multiplying the sum by $\dfrac{n-1}{2} = -\dfrac{1}{4}$, we obtain $-4'.22'',2$, and finally, adding this to $f'\left(a+\dfrac{1}{2}\right)$ and multiplying by $n = \dfrac{1}{2}$, we have $3^0.\ 18'.\ 43'',9$ to add to $f(a)$, and we obtain the same value as before, namely, $306^0.\ 43'.\ 45'',4$.

14. More convenient formulæ for interpolation are obtained, if Newton's formula of interpolation be so transformed, that in

it those differences solely are found, which stand on the same horizontal line, so that, setting out from the value of $f(a)$, the differences $f'\left(a+\frac{1}{2}\right)$, $f''(a)$, and $f'''\left(a+\frac{1}{2}\right)$, &c. are to be employed. The first two terms of the Newtonian formulæ can then be taken in.

But we have

$$f''(a+1)=f''(a)+f'''\left(a+\frac{1}{2}\right),$$

$$f'''\left(a+\frac{3}{2}\right)=f'''\left(a+\frac{1}{2}\right)+f^{\text{IV}}(a+1),$$

$$=f'''\left(a+\frac{1}{2}\right)+f^{\text{IV}}(a)+f^{\text{V}}\left(a+\frac{1}{2}\right),$$

$$f^{\text{IV}}(a+2)=f^{\text{IV}}(a+1)+f^{\text{V}}\left(a+\frac{3}{2}\right),$$

$$=f^{\text{IV}}(a)+2f^{\text{V}}\left(a+\frac{1}{2}\right)+f^{\text{VI}}(a+1),$$

$$f^{\text{V}}\left(a+\frac{5}{2}\right)=f^{\text{V}}\left(a+\frac{3}{2}\right)+f^{\text{VI}}(a+2)$$

$$=f^{\text{V}}\left(a+\frac{1}{2}\right)+f^{\text{VI}}(a+1)+f^{\text{VI}}(a+2),$$

&c.

Thus we obtain as the coefficient of $f''(a)$

$$\frac{n(n-1)}{1.2};$$

as the coefficient of $f'''\left(a+\frac{1}{2}\right)$

$$\frac{n(n-1)}{1.2}+\frac{n(n-1)(n-2)}{1.2.3}=\frac{(n+1)n(n-1)}{1.2.3},$$

as the coefficient of $f^{\text{IV}}(a)$

$$\frac{n(n-1)(n-2)}{1.2.3}+\frac{n.(n-1)(n-2)(n-3)}{1.2.3.4}$$

$$=\frac{(n+1)n.(n-1)(n-2)}{1.2.3.4},$$

and finally, as the coefficient of $f^{\text{v}}\left(a+\frac{1}{2}\right)$,

$$\frac{n\,(n-1)\,(n-2)}{1.2.3}+2.\frac{n\,(n-1)\,(n-2)\,(n-3)}{1.2.3.4}$$
$$+\frac{n\,(n-1)\,(n-2)\,(n-3)\,(n-4)}{1.2.3.4.5}=\frac{(n+2)\,(n+1)\,n\,(n-1)\,(n-2)}{1.2.3.4.5},$$

where the law of continuation is evident.

The whole formula is therefore

$$f(a+nw)=f(a)+nf'\left(a+\frac{1}{2}\right)+\frac{n\,(n-1)}{1.2}f''(a)$$
$$+\frac{(n+1)\,n\,(n-1)}{1.2.3}f'''\left(a+\frac{1}{2}\right)$$
$$+\frac{(n+1)\,n\,(n-1)\,(n-2)}{1.2.3.4}f^{\text{iv}}(a)$$
$$+\frac{(n+2)\,(n+1)\,n\,(n-1)\,(n-2)}{1.2.3.4.5}f^{\text{v}}\left(a+\frac{1}{2}\right)+\dots\quad(2).$$

If, in place of the differences which have $a+\frac{1}{2}$ under the sign of the function, we insert those which contain $a-\frac{1}{2}$, we have

$$f'\left(a+\frac{1}{2}\right)=f'\left(a-\frac{1}{2}\right)+f''(a),$$

$$f'''\left(a+\frac{1}{2}\right)=f'''\left(a-\frac{1}{2}\right)+f^{\text{iv}}(a),$$

$$f^{\text{v}}\left(a+\frac{1}{2}\right)=f^{\text{v}}\left(a-\frac{1}{2}\right)+f^{\text{vi}}(a).$$

These are the coefficients of the differences of odd orders, and on the other hand the coefficient of $f''(a)$ is

$$n+\frac{n\,(n-1)}{1.2}=\frac{n\,(n+1)}{2},$$

and that of $f^{\text{iv}}(a)$ is

$$\frac{(n+1)\,n\,(n-1)}{1.2.3}+\frac{(n+1)\,n\,(n-1)\,(n-2)}{1.2.3.4}=\frac{(n-1)\,n\,(n+1)\,(n+2)}{1.2.3.4}.$$

We obtain therefore

$$f(a+nw)=f(a)+nf'\left(a-\frac{1}{2}\right)+n\,.\,\frac{n+1}{2}\,.f''(a)$$

$$+\frac{(n-1)\,n\,(n+1)}{1\,.\,2\,.\,3}f'''\left(a-\frac{1}{2}\right)$$

$$+\frac{(n-1)\,n\,(n+1)\,(n+2)}{1\,.\,2\,.\,3\,.\,4}f^{\text{iv}}(a)$$

$$+\frac{(n-2)\,(n-1)\,n\,(n+1)\,(n+2)}{1\,.\,2\,.\,3\,.\,4\,.\,5}f^{\text{v}}\left(a-\frac{1}{2}\right)+\&c.$$

where again the law of continuation is evident.

If we assume now that it is required to interpolate a value whose argument lies between a and $a-w$, n is then negative. But if n always denote a positive value, we must employ $-n$ instead of n in the formula, and the formula immediately preceding will therefore be for this case:

$$f(a-nw)=f(a)-nf'\left(a-\frac{1}{2}\right)+n\,.\,\frac{n-1}{2}f''(a)$$

$$-\frac{(n-1)\,n\,(n+1)}{1\,.\,2\,.\,3}f'''\left(a-\frac{1}{2}\right)+\frac{(n+1)\,n\,(n-1)\,(n-2)}{1\,.\,2\,.\,3\,.\,4}f^{\text{iv}}(a)$$

$$-\frac{(n+2)\,(n+1)\,n\,(n-1)\,(n-2)}{1\,.\,2\,.\,3\,.\,4\,.\,5}f^{\text{v}}\left(a-\frac{1}{2}\right)+\ldots\ldots\ldots\ldots\ (3).$$

This formula is to be employed if interpolation is to be performed backwards. By writing again the two formulæ (2) and (3), as was before done with the Newtonian formulæ, we obtain

$$f(a+nw)=f(a)+n\left[f'\left(a+\frac{1}{2}\right)+\frac{n-1}{2}\left[f''(a),\right.\right.$$

$$+\frac{n+1}{3}\left[f'''\left(a+\frac{1}{2}\right)+\frac{n-2}{4}\left[f^{\text{iv}}(a)+\ldots\ldots\ldots\ldots\ (2a),\right.\right.$$

$$f(a-nw)=f(a)-n\left[f'\left(a-\frac{1}{2}\right)-\frac{n-1}{2}\left[f''(a),\right.\right.$$

$$-\frac{n+1}{3}\left[f'''\left(a-\frac{1}{2}\right)-\frac{n-2}{4}\left[f^{\text{IV}}(a)-\ldots\ldots\ldots\ldots\ (3a).\right.\right.$$

Imagine now a horizontal line to be drawn through the scheme of the functions and differences, in the neighbourhood of the place where the function for interpolation happens to lie; then, the first formula being employed if $a+nw$ be nearer to a than to $a+w$, and the second, if $a-nw$ lie nearer to a than to $a-w$, those differences must always be used which, on both sides of the horizontal line, lie nearest to the same. No farther attention is necessary to be paid to the signs of the differences, but every difference must be so corrected that it may approximate to that lying on the other side of the horizontal line.

For example, if the first formula be employed, and the argument therefore lie between a and $a+\frac{1}{2}w$, the horizontal line will between $f''(a)$ and $f''(a+1)$. We have then to add to $f''(a)$

$$+\frac{n+1}{3}f'''\left(a+\frac{1}{2}\right)=+\frac{n+1}{3}[f''(a+1)-f''(a)].$$

Thus if $f''(a)$ be $\left\{\begin{matrix}\text{less}\\\text{greater}\end{matrix}\right\}$ than $f''(a+1)$, the corrected $f''(a)$ will be $\left\{\begin{matrix}\text{greater}\\\text{less}\end{matrix}\right\}$ and therefore will always approximate to $f'''(a+1)$.

A somewhat greater degree of accuracy is attained if, for the last difference which is employed, the arithmetical mean of the two differences standing next to the horizontal line be taken. The arithmetical mean of the two differences will be denoted by means of the character of the difference function and the arithmetical mean of the two arguments which stand beneath, so that

$$f'(a+n)=\frac{f'\left(a+n-\frac{1}{2}\right)+f'\left(a+n+\frac{1}{2}\right)}{2}.$$

As before, there occur alternately fractions in the difference-characters of the even orders and whole numbers in those of the odd orders, so that no ambiguity can arise. If, for example, we terminate with the second difference, we must take, in the process of interpolation in the forward direction, the arithmetical mean of $f''(a)$ and $f''(a+1)$, that is, $f''\left(a+\frac{1}{2}\right)$. Then, instead of the term

$$\frac{n(n-1)}{1.2} f''(a)$$

we now make use of the term

$$\frac{n(n-1)}{1.2} f''\left(a+\frac{1}{2}\right),$$

that is, $$\frac{n(n-1)}{1.2}\left[f''(a)+\frac{1}{2}f'''\left(a+\frac{1}{2}\right)\right].$$

While then, by simply taking $f''(a)$, the omission of the whole of the third term is entailed, in the present case the only omission is that of

$$\left\{\frac{(n+1)\,n\,(n-1)}{1.2.3}-\frac{n(n-1)}{1.2.2}\right\}.f'''\left(a+\frac{1}{2}\right)$$

$$=\frac{n(n-1)\left(n-\frac{1}{2}\right)}{1.2.3} f'''\left(a+\frac{1}{2}\right).$$

For $n=\frac{1}{2}$, the error, as far as it depends on the third differences, is exactly nothing. For this case, when $n=\frac{1}{2}$, and the interpolation will therefore be made at the middle term, it is indifferent which of the two formulæ (2) or (3) is employed, since we may either set out with the argument a and interpolate forwards, or set out with the argument $a+w$ and interpolate backwards. The most convenient formula for this case is however obtained by a combination of the two. For $n=\frac{1}{2}$, the formula (2) becomes

$$f\left(a+\frac{1}{2}w\right)=f(a)+\frac{1}{2}f'\left(a+\frac{1}{2}\right)+\frac{\frac{1}{2}.-\frac{1}{2}}{1.2}f''(a)+\frac{\frac{3}{2}.\frac{1}{2}.-\frac{1}{2}}{1.2.3}f'''\left(a+\frac{1}{2}\right)$$

$$+\frac{\frac{3}{2}.\frac{1}{2}.-\frac{1}{2}.-\frac{3}{2}}{1.2.3.4}f^{\text{iv}}(a)+\ldots$$

On the contrary the formula (3) is, if we set out with the argument $a+w$,

$$f\left(a+\frac{1}{2}\right)=f(a+w)-\frac{1}{2}f'\left(a+\frac{1}{2}\right)+\frac{\frac{1}{2}\cdot-\frac{1}{2}}{1.2}f''(a+1)$$

$$-\frac{-\frac{1}{2}\cdot\frac{1}{2}\cdot\frac{3}{2}}{1.2.3}f'''\left(a+\frac{1}{2}\right)+\frac{\frac{3}{2}\cdot\frac{1}{2}\cdot-\frac{1}{2}\cdot-\frac{3}{2}}{1.2.3.4}f^{\text{IV}}(a+1).$$

If the arithmetical mean of the two formulæ be taken, all the terms in which occur differences of an odd order are got rid of, and we obtain for the interpolation the following very convenient formula, in which only the arithmetical means of the even orders appear:

$$f\left(a+\frac{1}{2}w\right)=f\left(a+\frac{1}{2}\right)-\frac{1}{8}f''\left(a+\frac{1}{2}\right)+\frac{3}{128}f^{\text{IV}}\left(a+\frac{1}{2}\right)$$

$$-\frac{5}{1024}f^{\text{VI}}\left(a+\frac{1}{2}\right)+\ldots\ldots \quad (4),$$

or $$f\left(a+\frac{1}{2}w\right)=f\left(a+\frac{1}{2}\right)-\frac{1}{8}\left[f''\left(a+\frac{1}{2}\right)-\frac{3}{16}\left[f^{\text{IV}}\left(a+\frac{1}{2}\right)\right.\right.$$

$$\left.\left.-\frac{5}{24}\left[f^{\text{VI}}\left(a+\frac{1}{2}\right)+\ldots\ldots \right.\right.\right. \quad (4a),$$

where the law of continuation is evident.

Example. Let the longitude of Mercury be required for January 4, 12h, then we must employ the formula (2a). The differences to be employed in this case will be the following:

	1st Diff.	2nd Diff.	3rd Diff.	4th Diff.
	7° . 0′ . 38″,0 +		2′ . 44″,3 +	
Jan. 4, 317°. 7′ . 29″,5 +		21′ . 32″,4 +		10″,1
	7 . 22 . 10 ,4		2 . 54 ,5	
6, 324 . 29 . 39 ,9		24 . 26 ,9		4 ,7

Here $n=\frac{1}{4}$, and so

$$\frac{n-1}{2}=-\frac{3}{8},\quad \frac{n+1}{3}=\frac{5}{12},\quad \frac{n-2}{4}=-\frac{7}{16};$$

paying no farther attention to the signs, we obtain

arithmetical mean of the 4th differences $\times \frac{7}{16} = 0°.\ 0'.\ 3'',2$

corrected 3rd difference $2'.51'',3 \times \frac{5}{12} = 1.11\ ,4$

corrected 2nd difference $22'.43'',8 \times \frac{3}{8} = 8.31\ ,4$

corrected 1st difference $7°.13'.39'',0 \times \frac{1}{4} = 1.48.24\ ,7$

Thus the true longitude for January 4·5 = 318°. 55′. 54″,2.

If the longitude be required for Jan. 5·5, we must employ the formula (3*a*), and take the differences which stand on both sides of the lower horizontal line. We find then the longitude for Jan. 5·5 = 322°. 36′. 56″,7.

For the purpose of employing the formula (4*a*), let the longitude be required for Jan. 5·0.

We obtain then

arithmetical mean of the 4th differences $\times - \frac{3}{16} = -\ 0°.0'.\ 1'',4$

corrected arith. mean of 2nd differences $\times - \frac{1}{8} = -\ 2.52\ ,3$

arith. mean of functions 320.48.34 ,7

whence the true longitude for

Jan. 5·0 = 320°. 45′. 42″,4.

Exhibiting now the differences of the interpolated values, we obtain

		1st Diff.	2nd Diff.	3rd Diff.
Jan. 4·0	317°. 7′. 29″,5			
		+1°. 48′. 24″,7		
4·5	318 . 55 . 54 ,2		+1′. 23″,5	
		1 . 49 . 48 ,2		+2″,6
5·0	320 . 45 . 42 ,4		1 . 26 ,1	
		1 . 51 . 14 ,3		2 ,8
5·5	322 . 36 . 56 ,7		1 . 28 ,9	
		1 . 52 . 43 ,2		
6·0	324 . 29 . 39 ,9			

The regular progression of the differences shews the correctness of the interpolation. This scrutiny by means of differences is made use of in all computations in which for certain arguments proceeding at small intervals a series of values of the function has been computed. If namely in any value, $f(a)$ for example, an error x has been committed, the scheme of the differences will be as follows:

$$
\begin{array}{lllll}
f(a-3w) & & & & \\
 & f'(a-\frac{5}{2}) & & & \\
f(a-2w) & & f''(a-2) & & \\
 & f'(a-\frac{3}{2}) & & f'''(a-\frac{3}{2})+x & \\
f(a-w) & & f''(a-1)+x & & f^{\text{IV}}(a-1)-4x \\
 & f'(a-\frac{1}{2})+x & & f'''(a-\frac{1}{2})-3x & \\
f(a)+x & & f''(a)-2x & & f^{\text{IV}}(a)+6x \\
 & f'(a+\frac{1}{2})-x & & f'''(a+\frac{1}{2})+3x & \\
f(a+w) & & f''(a+1)+x & & f^{\text{IV}}(a+1)-4x \\
 & f'(a+\frac{3}{2}) & & f'''(a+\frac{3}{2})-x & \\
f(a+2w) & & f''(a+2) & & \\
 & f'(a+\frac{5}{2}) & & & \\
f(a+3w) & & & &
\end{array}
$$

An error in the value of a function will thus exhibit itself very much magnified in the differences, and the greatest irregularity will be found in the horizontal line, in which the erroneous value of the function stands.

15. A case frequently occurs in which are employed the numerical values of the differential coefficients of a function whose analytical expression is not known, but only a series of numerical values of it, which follow each other at equal intervals, are given. In this case recourse must be had to the formula of interpolation for the computation of the numerical values of the differential coefficients.

By substituting, in the original formula (a) of No. 13, the values found for α, β, γ, δ, or, which is the same thing, developing by the Newtonian interpolation formula by powers of n, we have

$$
\begin{aligned}
f(a+nw) = f(a) &+ n\left[f'\left(a+\frac{1}{2}\right) - \frac{1}{2}f''(a+1) + \frac{1}{3}f'''\left(a+\frac{3}{2}\right) - \ldots\right] \\
&+ \frac{n^2}{1\,.\,2}\left[f''(a+1) - f'''\left(a+\frac{3}{2}\right) + \ldots\right] \\
&+ \frac{n^3}{1\,.\,2\,.\,3}\left[f'''\left(a+\frac{3}{2}\right) + \ldots\right.
\end{aligned}
$$

Now also since by Taylor's Theorem

$$
f(a+nw) = f(a) + \frac{df(a)}{da}nw + \frac{d^2f(a)}{da^2}\cdot\frac{n^2w^2}{1\,.\,2} + \frac{d^3f(a)}{da^3}\cdot\frac{n^3w^3}{1\,.\,2\,.\,3} + \ldots\ldots
$$

we obtain by comparison of the two series

$$\frac{df(a)}{da}=\frac{1}{w}\left[f'\left(a+\frac{1}{2}\right)-\frac{1}{2}f''(a+1)+\frac{1}{3}f'''\left(a+\frac{3}{2}\right)-\ldots.\right],$$

$$\frac{d^2f(a)}{da^2}=\frac{1}{w^2}\left[f''(a+1)-f'''\left(a+\frac{3}{2}\right)+\ldots.\right].$$

More convenient formulæ for the differential coefficients will be found from formula (2) in No. 14. Introducing in this formula the arithmetical mean of the odd differences, by making

$$f'\left(a+\frac{1}{2}\right)=f'(a)+\frac{1}{2}f''(a),$$

$$f'''\left(a+\frac{1}{2}\right)=f'''(a)+\frac{1}{2}f^{\text{IV}}(a),$$

we obtain

$$f(a+nw)=f(a)+nf'(a)+\frac{n^2}{1.2}f''(a)+\frac{(n+1)\,n.(n-1)}{1.2.3}f'''(a)$$
$$+\frac{(n+1)\,n^2\,(n-1)}{1.2.3.4}f^{\text{IV}}(a),$$

a formula which contains the even differences which stand with $f(a)$ on the same horizontal line, and on the contrary the arithmetical means of the odd differences, which lie on both sides of the horizontal line. Developing the same according to powers of n, we have

$$f(a+nw)=f(a)+n\left[f\ (a)-\frac{1}{6}f'''(a)+\frac{1}{30}f^{\text{V}}(a)-\frac{1}{140}f^{\text{VII}}(a)+\ldots\right]$$
$$+\frac{n^2}{1.2}\left[f''(a)-\frac{1}{12}f^{\text{IV}}(a)+\frac{1}{90}f^{\text{VI}}(a)-\ldots\right]$$
$$+\frac{n^3}{1.2.3}\left[f'''(a)-\frac{1}{4}f^{\text{V}}(a)+\frac{7}{120}f^{\text{VII}}(a)-\ldots\right]$$
$$+\frac{n^4}{1.2.3.4}\left[f^{\text{IV}}(a)-\frac{1}{6}f^{\text{VI}}(a)+\ldots\right]$$
$$+\frac{n^5}{1.2.3.4.5}\left[f^{\text{V}}(a)-\frac{1}{3}f^{\text{VII}}(a)+\ldots\right].$$

and from thence

$$\frac{df(a)}{da}=\frac{1}{w}\left[f'(a)-\frac{1}{6}f'''(a)+\frac{1}{30}f^{\text{v}}(a)-\frac{1}{140}f^{\text{vii}}(a)+\ldots\right],$$

$$\frac{d^2f(a)}{da^2}=\frac{1}{w^2}\left[f''(a)-\frac{1}{12}f^{\text{iv}}(a)+\frac{1}{90}f^{\text{vi}}(a)-\ldots\right],$$

$$\frac{d^3f(a)}{da^3}=\frac{1}{w^3}\left[f'''(a)-\frac{1}{4}f^{\text{v}}(a)+\frac{7}{120}f^{\text{vii}}(a)-\ldots\right],$$

&c. &c. .... (5).

If the differential coefficient be required for a function which is not found amongst those given, for example for $f(a+nw)$, we must put in this formula $a+n$ in place of a, so that

$$\frac{df(a+nw)}{da}=\frac{1}{w}\left[f'(a+n)-\frac{1}{6}f'''(a+n)+\frac{1}{30}f^{\text{v}}(a+n)+\ldots\right],$$

$$\frac{d^2f(a+nw)}{d^2a}=\frac{1}{w^2}\left[f''(a+n)-\frac{1}{12}f^{\text{iv}}(a+n)+\ldots\right],$$

&c. &c.....................(6).

The differences now to be employed do not appear in the scheme of the formulæ above, but must be first computed. For the even differences, for example $f''(a+n)$, this is easy, since they will be obtained by the ordinary interpolation formulæ, by considering $f''(a), f''(a+n)$, &c. as the functions, and the third differences as their first differences, &c. But the odd differences are arithmetical means, and a formula must first be developed for the interpolation of arithmetical means.

Now we have

$$f'(a+n)=\frac{f'\left(a+n-\frac{1}{2}\right)+f'\left(a+n+\frac{1}{2}\right)}{2}.$$

And by the interpolation formula (2) in No. 14,

$$f'\left(a-\frac{1}{2}+n\right)=f'\left(a-\frac{1}{2}\right)+nf''(a)+\frac{n.(n-1)}{1.2}f'''\left(a-\frac{1}{2}\right)$$
$$+\frac{(n+1)n.(n-1)}{1.2.3}f^{\text{iv}}(a)+\ldots$$

$$f'\left(a+\frac{1}{2}+n\right)=f'\left(a+\frac{1}{2}\right)+nf''(a)+\frac{n.(n+1)}{1.2}f'''\left(a+\frac{1}{2}\right)$$
$$+\frac{(n+1)\,n.(n-1)}{1.2.3}f^{\text{iv}}(a)+\&c.$$

We thus obtain, by taking the arithmetical mean of both formulæ, the formula for the interpolation of an arithmetical mean,

$$f'(a+n)=f'(a)+nf''(a)+\frac{n^2}{1.2}f'''(a)+\frac{1}{4}nf^{\text{iv}}(a)$$
$$+\frac{(n+1)\,n\,(n-1)}{1.2.3}f^{\text{iv}}(a)+\dots$$

The two terms

$$\frac{n^2}{1.2}f'''(a)+\frac{1}{4}nf^{\text{iv}}(a)$$

are composed of the arithmetical mean of the terms

$$\frac{n\,(n-1)}{1.2}f'''\left(a-\frac{1}{2}\right),$$

and

$$\frac{n\,(n+1)}{1.2}f'''\left(a+\frac{1}{2}\right),$$

which gives

$$\frac{n^2}{1.2}f'''(a)+\frac{n}{4}\left[f'''\left(a+\frac{1}{2}\right)-f'''\left(a-\frac{1}{2}\right)\right].$$

By combining the two terms which contain $f^{\text{iv}}(a)$, the formula above may be thus written:

$$f'(a+n)=f'(a)+nf''(a)+\frac{n^2}{2}f'''(a)+\frac{2n^3+n}{12}f^{\text{iv}}(a)+\dots\ (7).$$

By means of formulæ (5), (6), and (7), the numerical values of the differential coefficients of a function for any argument whatever may be computed from the even differences and the arithmetical means of the odd differences, when there is given a series of numerical values of the function following each other at equal intervals.

Other formulæ for the differential coefficients may now be developed, in which the simple differences of odd orders appear instead of the arithmetical means of the even differences.

In fact, if in the interpolation formula (3) the arithmetical mean of the even differences be introduced, by making

$$f(a)=f\left(a+\frac{1}{2}\right)-\frac{1}{2}f'\left(a+\frac{1}{2}\right),$$

$$f''(a)=f''\left(a+\frac{1}{2}\right)-\frac{1}{2}f'''\left(a+\frac{1}{2}\right),$$

$$f^{\text{iv}}(a)=f^{\text{iv}}\left(a+\frac{1}{2}\right)-\frac{1}{2}f^{\text{v}}\left(a+\frac{1}{2}\right),$$

we obtain, since

$$\frac{(n+1)\,n\,(n-1)}{1\,.\,2\,.\,3}-\frac{1}{2}\cdot\frac{n\,(n-1)}{1\,.\,2}=\frac{n\,.\,(n-1)\left(n-\frac{1}{2}\right)}{1\,.\,2\,.\,3},$$

$$f(a+nw)=f\left(a+\frac{1}{2}\right)+\left(n-\frac{1}{2}\right)f'\left(a+\frac{1}{2}\right)+\frac{n\,.\,(n-1)}{1\,.\,2}f''\left(a+\frac{1}{2}\right)$$

$$+\frac{n\,.\,(n-1)\left(n-\frac{1}{2}\right)}{1\,.\,2\,.\,3}f'''\left(a+\frac{1}{2}\right)+\frac{(n+1)\,n\,.\,(n-1)(n-2)}{1\,.\,2\,.\,3\,.\,4}f^{\text{iv}}\left(a+\frac{1}{2}\right)$$
$$+\ldots\ldots$$

If $n+\frac{1}{2}$ be written instead of n, the law of the coefficients will be simpler, since we obtain

$$f\left[a+\left(n+\frac{1}{2}\right)w\right]=f\left(a+\frac{1}{2}\right)+nf'\left(a+\frac{1}{2}\right)$$

$$+\frac{\left(n+\frac{1}{2}\right)\left(n-\frac{1}{2}\right)}{1\,.\,2}f''\left(a+\frac{1}{2}\right)+\frac{\left(n+\frac{1}{2}\right)n\,.\left(n-\frac{1}{2}\right)}{1\,.\,2\,.\,3}f'''\left(a+\frac{1}{2}\right)$$

$$+\frac{\left(n+\frac{3}{2}\right)\left(n+\frac{1}{2}\right)\left(n-\frac{1}{2}\right)\left(n-\frac{3}{2}\right)}{1\,.\,2\,.\,3\,.\,4}f^{\text{iv}}\left(a+\frac{1}{2}\right)+\ldots$$

Developing this formula according to powers of n, we obtain, since the terms independent of n are

$$f\left(a+\frac{1}{2}\right)-\frac{1}{8}f''\left(a+\frac{1}{2}\right)+\frac{3}{8\,.\,16}f^{\text{iv}}\left(a+\frac{1}{2}\right)-\ldots=f\left(a+\frac{1}{2}w\right),$$

$$f\left\{a+\left(n+\frac{1}{2}\right)w\right\}=f\left(a+\frac{1}{2}w\right)$$

$$+n\left[f'\left(a+\frac{1}{2}\right)-\frac{1}{24}f'''\left(a+\frac{1}{2}\right)+\frac{3}{640}f^{\text{v}}\left(a+\frac{1}{2}\right)-\dots\right]$$

$$+\frac{n^2}{1.2}\left[f''\left(a+\frac{1}{2}\right)+\frac{5}{24}f^{\text{iv}}\left(a+\frac{1}{2}\right)+\frac{259}{5760}f^{\text{vi}}\left(a+\frac{1}{2}\right)-\dots\right]$$

$$+\frac{n^3}{1.2.3}\left[f'''\left(a+\frac{1}{2}\right)-\frac{1}{8}f^{\text{v}}\left(a+\frac{1}{2}\right)+\frac{37}{1920}f^{\text{vii}}\left(a+\frac{1}{2}\right)-\dots\right]$$

$$+\frac{n^4}{1.2.3.4}\left[f^{\text{iv}}\left(a+\frac{1}{2}\right)-\frac{7}{24}f^{\text{vi}}\left(a+\frac{1}{2}\right)+\dots\right].$$

Comparing now this formula with the development of

$$f\left(a+\frac{1}{2}w+nw\right)$$

by Taylor's theorem, we find

$$\frac{df\left(a+\frac{1}{2}w\right)}{da}=\frac{1}{w}\left[f'\left(a+\frac{1}{2}\right)-\frac{1}{24}f'''\left(a+\frac{1}{2}\right)+\frac{3}{640}f^{\text{v}}\left(a+\frac{1}{2}\right)-\dots\right],$$

$$\frac{d^2f\left(a+\frac{1}{2}w\right)}{da^2}=\frac{1}{w^2}\left[f''\left(a+\frac{1}{2}\right)-\frac{5}{24}f^{\text{iv}}\left(a+\frac{1}{2}\right)+\frac{259}{5760}f^{\text{vi}}\left(a+\frac{1}{2}\right)-\dots\right]$$

&c. &c.............................(8).

This formula is used most conveniently, when it is required to compute the differential coefficients of a function for an argument which is the arithmetical mean of two consecutive arguments. For other arguments, for example,

$$a+\left(n+\frac{1}{2}\right)w,$$

we have again

$$\frac{df\left[a+\left(n+\frac{1}{2}\right)w\right]}{da}=f'\left(a+\frac{1}{2}+n\right)-\frac{1}{24}f'''\left(a+\frac{1}{2}+n\right)$$
$$+\frac{3}{640}f^{\text{v}}\left(a+\frac{1}{2}+n\right)+\ldots\ldots\ldots\ldots\ldots(9).$$

And here again the difference $f'\left(a+\frac{1}{2}+n\right)$, and in general all odd differences, will be computed by the ordinary interpolation formula. But since the even differences are arithmetical means, we obtain the formula to be employed for this from formula (7) for the interpolation of an arithmetical mean from the odd differences, by putting $a+\frac{1}{2}$ instead of a, and, for the purpose of finding $f''\left(a+\frac{1}{2}+n\right)$, increasing all the accents by unity, so that for example:

$$f''\left(a+\frac{1}{2}+n\right)=f''\left(a+\frac{1}{2}\right)+nf'''\left(a+\frac{1}{2}\right)+\frac{n^2}{2}f^{\text{iv}}\left(a+\frac{1}{2}\right)$$
$$+\frac{2n^3+n}{12}f^{\text{v}}\left(a+\frac{1}{2}\right)+\ldots$$

Example.

In the Berlin *Jahrbuch* for 1848, are the following Right Ascensions of the Moon:

	h.	h. m. s.	1st Diff. m. s.	2nd Diff. s.	3rd Diff. s.	4th Diff. s.
July 12	0	16.14.26,33				
			+ 25. 3,99			
	12	39.30,32		+ 23,75		
			25.27,74		− 1,39	
13	0	17. 4.58,06		22,36		− 0,85
			25.50,10		2,24	
	12	30.48,16		20,12		0,79
			26.10,22		3,03	
14	0	56.58,38		17,09		0,67
			26.27,31		3,70	
	12	18.23.25,69		13,39		
			26.40,70			
15	0	50. 6,39				

Supposing that from the above the first differential coefficients be required for July 13, 10^{h}, 11^{h}, and 12^{h}, and that formula (9) be employed, we must first compute the first and third differences for these times.

The third of the first differences corresponds to the argument July 13 6^h, and is $f'\left(a+\frac{1}{2}\right)$, and thus, for 10^h, 11^h, and 12^h, n is respectively $\frac{1}{3}$, $\frac{5}{12}$, and $\frac{1}{2}$.

By interpolating in the ordinary way, we obtain

	$f'\left(a+\frac{1}{2}+n\right)$,	$f''\left(a+\frac{1}{2}+n\right)$,
h.	m. s.	s.
for 10	+25.57,11	−2,51,
... 11	25.58,81	2,58,
... 12	26. 0,49	2,64.

From hence we obtain the differential coefficients

h.	m. s.
for 10	+25.57,21,
... 11	25.58,92,
... 12	26. 0,60,

the interval w being 12^h. If the same be required for one hour, we must divide by 12, and we then obtain the following values:

h.	m. s.
for 10	+2. 9,77,
... 11	2. 9,91,
... 12	2.10,05,

which express the hourly velocity of the moon in Right Ascension for those times.

If we had wished to employ formula (6), in which the arithmetical mean of the odd differences appears, we should have obtained, by taking $a =$ July 13 12^h (for 10^h for example, in which case $n = -\frac{1}{6}$,) according to formula (7):

$$f'\left(a-\frac{1}{6}\right) = +25'.56'',77 \text{ and } f'''\left(a-\frac{1}{6}\right) = -2'',51,$$

and from thence according to formula (6), for the differential coefficient, + 2.9,77.

The second differences are:

	h.	s.
for	10	+ 20,55,
	11	20,34,
	12	20,12,

Applying to these $-\frac{1}{12}$ part of the 4th differences and dividing by 144, we obtain the second differential coefficients for one hour as the unit:

	h.	s.
for	10	+ 0,1432,
	11	0,1417,
	12	0,1402.

Note. On the subject of Interpolations, the reader may compare the article by Encke in the *Jahrbuch* for 1830, and the above-mentioned article on Mechanical Quadratures in the *Jahrbuch* for 1837.

SPHERICAL ASTRONOMY.

FIRST SECTION.

THE VISIBLE SPHERE OF THE HEAVENS, AND ITS DAILY MOTION.

In Spherical Astronomy are treated of the places of the bodies in the visible sphere of the heavens, by referring them by means of spherical co-ordinates to certain great circles devised on the sphere. Spherical Astronomy gives then the means both of determining the place of a heavenly body in relation to these great circles, and also the relative positions of these circles with regard to each other. It is necessary therefore first to obtain a knowledge of these great circles, whose planes are the fundamental planes of the different systems of co-ordinates, and, at the same time, of the means which have been employed to reduce the place of a body, which is given for one of these fundamental planes, to another system of co-ordinates.

Some of these co-ordinates are independent of the daily motion of the sphere, others, on the contrary, are referred to planes, which do not partake of this daily motion. The heavenly bodies will therefore, if they are referred to these latter planes, constantly change their places, and it will be of importance to obtain a knowledge of these changes and of the phenomena that are consequent upon them. But, since the heavenly bodies besides these motions which are common to all, exhibit also other, and those much slower motions, by reason of which they change their places also with respect to the systems of co-ordinates which are independent of the daily motion, it is not sufficient to determine solely the position of a heavenly body, but it is besides necessary that the time corresponding to this position should be given. It is therefore necessary to obtain a knowledge of the methods by which the daily motion of the sphere, partly by itself and partly in connexion with the motion of the sun in the same, is made to serve as a measure of time.

I. *On the different systems of Planes and Circles on the visible Sphere of the Heavens.*

1. The heavens appear to us as the hollow surface of a sphere, on which we see the stars projected and at the centre of which we are situated. For the purpose of determining the positions of the heavenly bodies on this visible sphere various systems of spherical co-ordinates have been devised. The first of these systems is that of Azimuths and Altitudes. In this system the fundamental plane is the plane of the horizon, which is determined by the surface of a fluid at rest, supposed to be extended to infinity. This plane intersects the visible sphere of the heavens in a great circle, which is called *the horizon*. The plane of the horizon may also be defined as that plane which is perpendicular to the plumbline, that is, to the direction of gravity at the surface of the earth. The plumb-line itself cuts the visible sphere of the heavens in two points, which are the poles of the horizon, and of which the one situated above the horizon is called the *zenith*, and the opposite one the *nadir*.

By means of the zenith-point and the horizon, the position of a heavenly body on the sphere of the heavens can now be determined. A great circle for instance is drawn through the zenith and the body whose position is to be determined, and this circle is consequently perpendicular to the horizon. By determining now the point of intersection of this circle and the horizon, and by reckoning from this point upwards along the great circle the number of degrees between the horizon and the body, and in like manner along the horizon the number of degrees to a certain origin or zero of co-ordinates, we have the two spherical co-ordinates, by means of which the position of the body is to be determined. The great circle passing through the zenith and the object is called a *vertical circle;* the arc of this circle between the horizon and the object is called the *altitude of the body;* and the arc between the object and the zenith is called its *zenith-distance.* The sum of the altitude and zenith-distance is always equal to 90°. The arc of the horizon between the vertical circle passing through the object and a point arbitrarily chosen is called *its azimuth.* This zero-point is taken arbitrarily, but for simplicity let it be chosen so that it may coincide with the zero-point of the second system of co-ordinates under consideration. The direction in which azimuths are reckoned is likewise unimportant; let us take therefore, as in the second system of co-ordinates, the direction of the diurnal motion, in which we reckon round from left to right from 0° to 360°. Small circles which are parallel to the horizon, are named also *horizontal circles* or *almucantarats.*

We can, instead of using these spherical co-ordinates, determine the place of an object by means of rectangular co-ordinates, referred to a system of axes of which the axis of z is perpendicular to the plane of the horizon, whilst the axes of x and y are situated in that plane, and so that the axis of x is drawn towards the zero-point of azimuths, the positive axis of y towards an azimuth of 90°. Indicating then the azimuth by A, the altitude by h, we have

$$x = \cos h \cos A, \quad y = \cos h \sin A, \quad z = \sin h.$$

NOTE. In order to observe these co-ordinates we have an instrument completely corresponding to this co-ordinate system, namely the altitude and azimuth instrument. This consists essentially of an horizontal divided circle, which stands upon three adjusting footscrews, and which can be placed horizontal by means of a spirit-level. This circle represents the plane of the horizon. In the centre of this stands a vertical pillar, which carries a second circle parallel to it, and therefore likewise perpendicular to the horizon. About the centre of this vertical circle revolves a telescope which is connected with a pointer, by means of which we can observe the direction of the telescope with reference to the circle. The vertical pillar carries likewise a pointer at right angles to it, which indicates the azimuth upon the horizontal circle. If now it be known what points upon the circles correspond to the zero of the azimuth and to the zenith-point, we are enabled by means of such an instrument, if we direct the telescope to an object, to find its azimuth and its altitude or zenith-distance. Besides this there are still other instruments, with which altitudes only can be observed. They are called altitude instruments; and are either quadrants, sextants, or entire circles. Instruments, with which we observe azimuths only, are called theodolites.

2. Heavenly bodies change their places on the visible sphere of the heavens, and indeed every star, in consequence of the daily motion of the earth in that period of time which is called a sidereal day, describes a circle on the sphere which generally is a small circle. Since the planes of all these circles are parallel to one another, they are called *parallel circles.* The axis about which the daily motion takes place is called the axis of the earth. It meets the visible sphere of the heavens in two points, of which that one which is visible on the northern hemisphere of the earth is called the *north pole,* while the other is called the *south pole.* The great circle, the poles of which are those two poles of the earth, is called the *equator.* The position of the earth's axis can be determined by means of the first system of co-ordinates, by finding the azimuth and the altitude of the pole. This last quantity is called the *latitude* of the place of observation; and it is equal to the angle between the equator and the zenith of the place. The complement of the elevation of the pole to 90° is the altitude of the equator above the horizon or the altitude of the equator at the place of observation.

The daily motion of the heavenly bodies now serves for the foundation of a second system of co-ordinates. Great circles which pass through the star and the poles, and are therefore perpendicular to the equator, are called *declination-circles,* or *hour-circles.* The arc of such a great circle which is included between the equator and the star, is called the *declination* of the star; on the contrary, the arc between the north pole and the star is called the *polar-distance.* The declination is said to be positive, when the star lies in that part of the declination-circle which is included between the equator and the north pole; negative, when the star is in that part which is between the equator and the south pole.

Declination and polar-distance are always complementary to each other, and correspond to the altitude and zenith-distance in the first system of co-ordinates. Analogous to the azimuth there is the hour-angle, that is, the angle formed by the hour-circle passing through the star and a determinate circle taken as the origin, and it is reckoned in the direction of the daily motion from 0° to 360°. For the first hour-circle, that has been chosen which passes through the zenith, and is called the *meridian*. This meets the horizon in two points, of which that one which lies between the north pole and the equator is called the north point, and the other for a similar reason the south point. The latter is the zero from which the azimuths are reckoned. Ninety degrees from the north and south points lie the west and east points, which are at the same time the points of intersection of the horizon and equator.

Instead of determining the place of an object by the two spherical co-ordinates, declination and hour-angle, it can be determined by means of rectangular co-ordinates, by referring it to three co-ordinate axes, of which the positive axis of z is drawn perpendicular to the equator and towards the north pole, while the axes of x and y lie in the plane of the equator, and so that the positive axis of x is drawn towards the zero-point, the positive axis of y, on the contrary, to the ninetieth degree of the hour-angle. Designating then the declination by δ, the hour-angle by t, we have

$$x' = \cos\delta \cos t, \quad y' = \cos\delta \sin t, \quad z' = \sin\delta.$$

NOTE.—Corresponding to this second system of co-ordinates, declination and hour-angle, there is another kind of instrument called the parallactic instrument or equatoreal. In this the circle, which, in the first kind of instruments, is parallel to the horizon, is parallel to the equator, so that the pillar at right-angles to it is in the direction of the earth's axis. The circle then which is parallel to this pillar is an hour-circle, and the angles which are read off on the first circle or that parallel to the equator, are hour-angles. Therefore if we know those points of the circle, which correspond to the zero of hour-angle, and to the pole, we can, by means of such an instrument find the declinations and hour-angles of celestial objects.

3. In this second system, one of the spherical co-ordinates, viz. the declination, is constant, on the other hand the hour-angle changes at every instant, since we begin to reckon it from a point of the heavens which does not partake of the daily motion.

In order now to have the second co-ordinate also constant, we choose as zero-point a determinate point of the equator, and naturally that one in which the equator is cut by the fundamental plane of the following system of co-ordinates, viz. the ecliptic. By virtue of the annual motion of the earth about the sun, the centre of the sun appears to describe in the course of a year a great circle in the heavens, which is called the *ecliptic* or sun's path. This great circle is inclined to the equator at an angle of about $23\frac{1}{2}^0$, which is called the *obliquity of the ecliptic.* The points of intersection of the ecliptic with the equator are called respectively the *vernal* and *autumnal equinoxes,* because when the sun, on the 21st March and 23rd September in each year, passes through these points, the days and nights at all points of the earth are equal*. The points of the ecliptic which are 90^0 distant from the equinoxes are called the *solstitial points.*

This newly introduced co-ordinate is reckoned on the equator from the vernal equinox, and is called the right ascension of the star. It is reckoned from west to east from 0^0 to 360^0, that is in the direction opposite to that of the daily motion. Instead of the spherical co-ordinates of right ascension and declination we can introduce as before rectangular co-ordinates, by referring the place of a body to three rectangular axes, of which the positive axis of z is drawn perpendicular to the equator and towards the north pole, while the axes of x and y lie in the plane of the equator, and so that the positive axis of x is directed towards the zero-point, the positive axis of y to a point having 90^0 of right ascension. Putting then α for the right ascension, we have

$$x'' = \cos\delta\cos\alpha, \quad y'' = \cos\delta\sin\alpha, \quad z'' = \sin\delta.$$

The co-ordinates α and δ are constant for every star, but in order to obtain thereby the place of a star on the visible

* Since, namely, the sun at that time is in the equator, and since the equator and horizon as great circles bisect each other, it follows that the sun on these days is as long above as he is below the horizon.

sphere of the heavens for any particular instant, we must know the position of the vernal equinox in the heavens for that instant. Sidereal time determines the place of the equinox, which is equal to its hour-angle, and of which 24 hours are reckoned to a sidereal day. It is 0^h sidereal when the hour-angle of the vernal equinox is zero and therefore the vernal equinox is on the meridian; it is 1^h sidereal when the hour-angle of the vernal equinox is one 24th part of the circumference or 15°; &c. This is the reason why the equator besides being divided into 360° is also divided into 24^h. Putting then θ for sidereal time, we have always

$$\theta - t = \alpha, \text{ and therefore } t = \theta - \alpha.$$

If, for example, the right ascension $= 190^\circ.20'$, the sidereal time $\theta = 4^h$, then $t = 229^\circ.40'$.

From the equation for t, we have when $t = 0$, $\theta = \alpha$. Every star therefore comes to the meridian or culminates at a sidereal time which expresses also its right ascension in time*. If we know therefore, the right ascension of an object, which at a given instant is on the meridian, we have the sidereal time corresponding to that instant.

NOTE. The co-ordinates of the third system we can find by means of instruments of the second description, if we know the sidereal time. In any particular case the co-ordinates may be determined also by means of instruments of the first description,

* The conversion of arcs into time and *vice versâ* is of very frequent occurrence. To convert arc into time we must divide by 15; we have then

$$(15a + b) \text{ degrees, } (15c + d) \text{ minutes, } (15e + f) \text{ seconds,}$$

which is in time,

$$a \text{ hours, } 4b + c \text{ minutes, } 4d + e \text{ and } \frac{f}{15} \text{ seconds.}$$

$$\text{For Example } 239^\circ.18'.46'',75 = 15^h, \overline{14 \times 4 + 1^m}, \overline{3 \times 4 + 3^s} + 0,117^s$$
$$= 15^h.57^m.15^s, 117.$$

On the other hand, to convert time into degrees we must multiply by 15 and afterwards divide the minutes and seconds by 60. We then have

$$a \text{ hours, } 4b + c \text{ minutes, } 4d + e \text{ seconds}$$

to convert into degrees, when it becomes

$$15a + b \text{ degrees, } 15c + d \text{ minutes, and } 15e \text{ seconds.}$$

Thus,

$$15^h.57^m.15^s.117 = \overline{225 + 14} \text{ degrees,}$$
$$15 + 3 \text{ minutes and } 46,75 \text{ seconds,}$$
$$= 239^\circ.18'.46'',75.$$

namely, by transits of stars over the meridian, since right ascensions are determined by observation of transits and declinations by observing the altitude of the star on the meridian, if the elevation of the equator or of the pole at the place of observation be known. The meridian circle which is an altitude circle placed in the plane of the meridian serves for these observations. Should the instrument not serve for determination of altitude, but simply for observations of transits of stars over the meridian, it is then simply an azimuthal instrument which is placed in the plane of the meridian, and is therefore called a *transit instrument.* By observing with such an instrument, together with a good clock, the transits of stars over the meridian, their differences of right ascension are found. Since the zero-point of right ascensions cannot be observed directly, it is somewhat more difficult to determine absolute right ascensions.

4. The fourth system of co-ordinates is that one of which the ecliptic is the fundamental plane. Great circles which pass through the poles of the ecliptic, and so are perpendicular to it, are called *circles of latitude;* and the arc of such a circle which is included between the ecliptic and the star, is called the latitude of the star. This is positive, when the star is in the northern of the two hemispheres formed by the ecliptic, and negative when the star is in the southern hemisphere. The other co-ordinate, the longitude, is reckoned on the ecliptic, and is the arc between the circle of latitude of the star and the vernal equinox. It is reckoned round from 0° to 360° in the same way as the right ascension, and therefore contrary to the daily motion of the heavens*. The circle of latitude, of which the longitude is zero, is called the *equinoctial colure.* On the other hand, that circle, of which the longitude is 90°, is called the solstitial colure. The arc of this colure which is included between the equator and the ecliptic, is the *obliquity of the ecliptic;* this is also equal to the arc of the great circle between the poles of the equator and of the ecliptic.

The longitude is always, in the following pages, represented by λ, the latitude by β, the obliquity by ϵ.

* The longitudes of stars are also often indicated by the signs of the zodiac, each of which includes 30°.

Thus 6 signs $15^\circ = 195^\circ$ of longitude.

If we express the spherical co-ordinates β and λ by means of a system of three rectangular axes, of which the positive axis of z is drawn perpendicular to the ecliptic and towards the north pole, whilst the axes of x and y lie in the plane of the ecliptic, and so that the positive axis of x is drawn towards the zero-point, the positive axis of y to a point having 90^0 of longitude, we have:

$$x''' = \cos\beta\cos\lambda,\quad y''' = \cos\beta\sin\lambda,\quad z''' = \sin\beta.$$

NOTE. Formerly there were instruments by which the spherical co-ordinates of the several systems could be observed, and therefore also the longitudes and latitudes. At present these are no longer in use. The co-ordinates of longitude and latitude are never determined by direct observation, but are always obtained by means of computation from co-ordinates of the other systems.

II. *On the transformation from one to another of the several systems of Co-ordinates.*

5. In order to reduce the place of a star, which has been referred to the co-ordinate system of azimuth and altitude to the co-ordinate system of hour-angle and declination, we have only to turn the axis of z in the first system in the plane of x and z, reckoning according to the positive directions of x and z, through an angle $90^0 - \phi$ (where ϕ represents the elevation of the pole), since the axis of y is common to both systems, and we then obtain from the formula (1a) for the transformation of co-ordinates or also by the formulæ of spherical trigonometry, in the triangle between the zenith, the pole, and the star*:

$$\sin\delta = \sin\phi\sin h - \cos\phi\cos h\cos A,$$
$$\cos\delta\sin t = \cos h\sin A,$$
$$\cos\delta\cos t = \sin h\cos\phi + \cos h\sin\phi\cos A.$$

Should we desire to have the formulæ in a form more adapted for logarithmic computation, we must put

$$\sin h = m\cos M,$$
$$\cos h\cos A = m\sin M,$$

* The three sides of this triangle are respectively,

$$90^0 - h,\ 90^0 - \delta,\ 90^0 - \phi,$$

and these stand opposite to the angles

$$t, 180^0 - A, \text{ and the angle of the star.}$$

from which we obtain

$$\sin\delta = m\sin(\phi - M),$$
$$\cos\delta\sin t = \cos h\sin A,$$
$$\cos\delta\cos t = m\cos(\phi - M).$$

These formulæ give the required results without any ambiguity. Since then every part is found by means of sines and cosines, we have only to observe the proper sign in order to determine the right quadrant for the required part. The subsidiary angles which have been introduced for the transformation of such formulæ have always a geometrical meaning, which is found easily in every case. Geometrically considered, the introduction of the subsidiary angle depends upon these circumstances, either that we divide the oblique-angled spherical triangle into two right-angled triangles, or compose it out of two right-angled triangles. In the present case we must imagine a perpendicular let fall from the star on the opposite side $90^0 - \phi$, or that side produced, and then

$$\tan h = \cos A \operatorname{cotan} M$$

where, according to the third of the formulæ (12) in No. 8 of the Introduction, M is the arc between the zenith and the foot of the perpendicular; and again, according to the first of the formulæ (12), m is the cosine of the same perpendicular, and thus

$$\sin h = \cos P \cos M,$$

if we put P for the perpendicular.

Let ϕ (the elevation of the pole) $= 52^0.30'.16'',00$,
$h = 16^0.11'.44'',0$, $A = 202^0.4'.15'',5$.

The computation will then be as follows:

$\cos A$	$-$ 9·9669481,	$m\sin M$	$-$ 9·9493620,
$\cos h$	9·9824139,	$m\cos M$	9·4454744,
$\sin A$	$-$ 9·5749045,		
		$M =$	$-72^0.35'.54'',61$,
		$\sin M$	$-$ 9·9796542,

$\phi - M = 125^\circ . 6' . 10'',61$

$\sin(\phi - M)$	$9{\cdot}9128171,$	$\cos\delta \sin t - 9{\cdot}5573184,$
m	$9{\cdot}9697078,$	$\cos\delta \cos t - 9{\cdot}7284114,$
$\sin\delta$	$9{\cdot}8825249,$	
$\cos\delta$	$9{\cdot}8104999,$	

$t = 223^\circ . 56' . 2'',22, \qquad \delta = +49^\circ . 43' . 46'',00$

$\cos(\phi - M) - 9{\cdot}7597036 \quad \cos t - 9{\cdot}9189115.$

6. Far more frequently, however, the converse case to this is used, where it is required to reduce a position which has been referred to the co-ordinate system of hour-angle and declination to the co-ordinate system of azimuth and altitude. We have then again, according to formula (1), for the transformation of co-ordinates, the following equations:

$$\sin h = \sin\phi \sin\delta + \cos\phi \cos\delta \cos t,$$
$$\cos h \sin A = \cos\delta \sin t,$$
$$\cos h \cos A = -\cos\phi \sin\delta + \sin\phi \cos\delta \cos t,$$

which we can again give in a more convenient form by introducing a subsidiary angle.

If we put namely:

$$\cos\delta \cos t = m \cos M,$$
$$\sin\delta = m \sin M,$$

then

$$\sin h = m \cos(\phi - M),$$
$$\cos h \sin A = \cos\delta \sin t,$$
$$\cos h \cos A = m \sin(\phi - M);$$

or also

$$\tan A = \frac{\cos M \tan t}{\sin(\phi - M)},$$

$$\tan h = \frac{\cos A}{\tan(\phi - M)}\text{*}.$$

* Since the azimuth lies always upon the same side of the meridian as the hour-angle, we can also by application of these last formulæ never be in doubt as to the quadrant in which it is to be taken.

If the zenith-distance alone be required, the following formulæ are convenient,

$$\cos z = \cos(\phi - \delta) - 2\cos\phi\cos\delta\sin^2\frac{t}{2},$$

or

$$\sin^2\frac{z}{2} = \sin^2\frac{(\phi-\delta)}{2} + \cos\phi\cos\delta\sin^2\frac{t}{2}.$$

If we now put

$$n = \sin\frac{\phi-\delta}{2},$$

$$m = \sqrt{\cos\phi\cos\delta},$$

then

$$\sin^2\frac{z}{2} = n^2\left(1 + \frac{m^2}{n^2}\sin^2\frac{t}{2}\right),$$

or, if we put

$$\frac{m}{n}\sin\frac{t}{2} = \tan\lambda,$$

$$\sin\frac{z}{2} = \frac{n}{\cos\lambda}.$$

If $\sin\lambda$ be greater than $\cos\lambda$, then the formula

$$\sin\frac{z}{2} = \frac{m}{\sin\lambda}\sin\frac{t}{2}$$

is preferable for computation.

In the formula for computation of n we must moreover, as we shall see further on, use $\phi - \delta$ for stars which culminate south of the zenith, and $\delta - \phi$ for stars which culminate north of the zenith.

Employing now the Gaussian formulæ for the triangle between the zenith, the pole, and the star, we have, if we denote the angle at the star by p,

$$\cos\frac{z}{2}\sin\frac{A-p}{2} = \sin\frac{t}{2}\sin\frac{\phi+\delta}{2},$$

$$\cos\frac{z}{2}\cos\frac{A-p}{2} = \cos\frac{t}{2}\cos\frac{\phi-\delta}{2},$$

$$\sin\frac{z}{2}\sin\frac{A+p}{2}=\sin\frac{t}{2}\cos\frac{\phi+\delta}{2},$$

$$\sin\frac{z}{2}\cos\frac{A+p}{2}=\cos\frac{t}{2}\sin\frac{\phi-\delta}{2}.$$

Reckoning now the azimuth from the north point, as would be done in the case of the pole star, we must put $180^\circ-A$ for A in these formulæ, and we obtain

$$\cos\frac{z}{2}\sin\frac{p+A}{2}=\cos\frac{t}{2}\cos\frac{\delta-\phi}{2},$$

$$\cos\frac{z}{2}\cos\frac{p+A}{2}=\sin\frac{t}{2}\sin\frac{\delta+\phi}{2},$$

$$\sin\frac{z}{2}\sin\frac{p-A}{2}=\cos\frac{t}{2}\sin\frac{\delta-\phi}{2},$$

$$\sin\frac{z}{2}\cos\frac{p-A}{2}=\sin\frac{t}{2}\cos\frac{\delta+\phi}{2}.$$

The case often occurs in which we have for a given latitude to make a great many such transformations, and for which, for the sake of more ready computation, tables are constructed beforehand*. For this case the second transformation, which is given in No. 6 of the Introduction, for the three fundamental equations, is especially convenient. We easily obtain formulæ suitable for the present case, by putting respectively in the equations there given

$$90^\circ-h,\quad 90^\circ-\delta,\quad 90^\circ-\phi,\quad 180^\circ-A,\quad \text{and } t,$$

instead of

$$a,\qquad b,\qquad c,\qquad B,\qquad \text{and } A.$$

For the sake of perspicuity we will now repeat these transformations together with the equations now arrived at: the equations are

* If, for example, we wish to obtain the places of stars of which the positions are given in right ascension and declination, by means of an instrument on which we can only read altitudes and azimuths, it will then be previously necessary to compute the hour-angle from the right ascension and the sidereal time.

$$(a)\qquad \sin h = \sin\phi \sin\delta + \cos\phi \cos\delta \cos t,$$
$$(b)\qquad \cos h \sin A = \cos\delta \sin t,$$
$$(c)\qquad \cos h \cos A = -\cos\phi \sin\delta + \sin\phi \cos\delta \cos t.$$

Denoting then by A_0 and δ_0 these values of A and δ, which, when substituted in the previous equations, give $h = 0$, we have

$$(d)\qquad 0 = \sin\phi \sin\delta_0 + \cos\phi \cos\delta_0 \cos t,$$
$$(e)\qquad \sin A_0 = \cos\delta_0 \sin t,$$
$$(f)\qquad \cos A_0 = -\cos\phi \sin\delta_0 + \sin\phi \cos\delta_0 \cos t.$$

Multiplying (f) by $\cos\phi$ and subtracting from it equation (d) after first multiplying it by $\sin\phi$, and, in addition, multiplying equation (f) by $\sin\phi$ and adding to it equation (d), previously multiplied by $\cos\phi$, we obtain

$$\left.\begin{aligned} \cos A_0 \cos\phi &= -\sin\delta_0 \\ \cos A_0 \sin\phi &= \cos\delta_0 \cos t \\ \sin A_0 &= \cos\delta_0 \sin t \end{aligned}\right\} \dots\dots\dots\dots\dots (A).$$

Put then

$$\left.\begin{aligned} \sin\phi &= \sin\gamma \cos B, \\ \cos\phi \cos t &= \sin\gamma \sin B, \\ \cos\phi \sin t &= \cos\gamma, \end{aligned}\right\} \dots\dots\dots\dots\dots (B),$$

we thus obtain from equation (d)

$$0 = \sin\gamma \sin(\delta_0 + B)$$

or

$$\delta_0 = -B,$$

and from (a)

$$\sin h = \sin\gamma \sin(\delta + B).$$

In addition, we obtain, by subtracting the product of equations (c) and (e) from the product of equations (b) and (f),

$$\cos h \sin(A - A_0) = \cos\phi \sin t \sin(\delta - \delta_0) = \cos\gamma \sin(\delta + B),$$

and similarly, if we add the product of equations (b) and (e) and that of equations (a) and (d) to the product of equations (c) and (f),

$$\cos h \cos(A - A_0) = \cos\delta \cos\delta_0 \sin^2 t + \sin\delta \sin\delta_0 + \cos\delta \cos\delta_0 \cos^2 t,$$
$$= \cos(\delta - \delta_0) = \cos(\delta + B).$$

This system of formulæ is thus when perfectly formed

$$\left.\begin{aligned}\sin\phi &= \sin\gamma\cos B\\ \cos\phi\cos t &= \sin\gamma\sin B\\ \cos\phi\sin t &= \cos\gamma\end{aligned}\right\}\ldots\ldots\ldots\ldots\ldots (1),$$

$$\left.\begin{aligned}\sin B &= \cos A_0\cos\phi\\ \cos B\cos t &= \cos A_0\sin\phi\\ \cos B\sin t &= \sin A_0\end{aligned}\right\}\ldots\ldots\ldots\ldots\ldots (2),$$

$$\left.\begin{aligned}\sin h &= \sin\gamma\sin(\delta+B)\\ \cos h\cos(A-A_0) &= \cos(\delta+B)\\ \cos h\sin(A-A_0) &= \cos\gamma\sin(\delta+B)\end{aligned}\right\}\ldots\ldots\ldots\ldots\ldots (3).$$

Putting $D=\sin\gamma,\ C=\cos\gamma,\ A-A_0=u,$

these formulæ become

$$\begin{aligned}\tan B &= \cot\phi\cos t,\\ \tan A_0 &= \sin\phi\tan t,\\ \sin h &= D\sin(B+\delta),\\ \tan u &= C\tan(B+\delta),\\ A &= A_0+u,\end{aligned}$$

and D and C are then the sine and cosine of an angle γ, which is given by the equation

$$\cot\gamma=\sin B\tan t=\cot\phi\sin A_0{}^{*}.$$

These are the formulæ communicated by Gauss in "*Schumacher's Auxiliary Tables lately edited by Warnstorff*, p. 135, &c." Arranging now the quantities D, C, B and A in Tables, of which the argument is t, the computation of the altitude and azimuth from the hour-angle and the declination, is reduced to the computation of the preceding formulæ:

$$\begin{aligned}\sin h &= D\sin(B\mp\delta),\\ \tan u &= C\tan(B+\delta),\\ A &= A_0+u.\end{aligned}$$

* We have in fact,

$$\begin{aligned}\cot\phi\sin A_0 &= -\sin\delta_0\tan t\\ &= \sin B\tan t.\end{aligned}$$

In Warnstorff's Auxiliary Tables we find such a table calculated for the latitude of the Altona Observatory. In addition, it is only necessary to calculate these tables from $t = 0$ to $t = 6^h$. Then it follows from the equation $\tan A_0 = \sin \phi \tan t$, that A_0 and t lie always in the same quadrant, and that for an hour-angle $= 12^h - t$ we have only to take $180^0 - A$. It further results from the equation for B that this angle is negative when t is $> 6^h$ or $> 90^\circ$, and that we must use the value $- B$ for an hour-angle $= 12^h - t$. The quantities

$$C = \cos \phi \sin t, \text{ and } D^2 = \sin^2 \phi + \cos^2 \phi \cos^2 t,$$

are on the contrary not altered, if we put $180^0 - t$ instead of t in these expressions. If t lies between 12^h and 24^h, we have to continue the calculation with the complement of t to 24^h, and to take afterwards for the determination of A its complement to 360°.

It is now easy to find the geometrical meaning of the subsidiary angles. Since δ_0 is that value of δ which makes $h = 0$ in the first of the primary equations, therefore δ_0 is the declination of the point in which the hour-circle passing through the star cuts the horizon, and similarly A_0 is the azimuth of that point. Further, since $B = -\delta_0$, $B + \delta$ is the arc SF, (fig. 1*) of the hour-circle produced to the horizon. Considering then the right-angled triangle FOK, which is formed by the horizon, the equator, and the side $FK = B$, we have from the sixth of the formulæ (12) in the Introduction, since the angle at O is $90^0 - \phi$,

$$\sin \phi = \cos B \sin OFK.$$

But since also $\sin \phi = D \cos B$, D is the sine and consequently C the cosine of the angle OFK. Lastly, as is easy to perceive, the arc $FH = A_0$ and the arc $FG = u$.

We thus find the formulæ previously given, through the consideration of the three right-angled triangles PFH, OFK, and SFG.

* In this figure P is the pole, Z the zenith, OH the horizon, OA the equator, and S the star.

The first triangle gives

$$\tan A_0 = \tan t \sin \phi ;$$

the second

$$\tan B = \cot \phi \cos t,$$

$$\cot \gamma = \sin B \tan t = \cot \phi \sin A_0 ;$$

and lastly, the third

$$\sin h = \sin \gamma \sin (B + \delta),$$

$$\tan u = \cos \gamma \tan (B + \delta).$$

The same subsidiary quantities can now also be employed for the solution of the converse problem treated of in No. 5, to calculate from the altitude and azimuth of a star its hour-angle and declination. We have namely in the right-angled triangle SLK, if we put $LG = B$, $LK = u$, $AL = A_0$ and the cosine of the angle $SLK = C$ and the sine $= D$,

$$C \tan (h - B) = \tan u,$$

$$D \sin (h - B) = \sin \delta,$$

and

$$t = A_0 - u ;$$

where now

$$\tan B = \cot \phi \cos A,$$

$$\tan A_0 = \sin \phi \tan A,$$

and D and C are the sine and cosine respectively of an angle γ, which is given by the equation

$$\cot \gamma = \sin B \tan A.$$

We have also for the calculation of the subsidiary quantities the same formulæ as before, with this difference only, that everywhere A occurs instead of t, and hence we can employ the same Auxiliary Tables as before, if we only now take as argument the azimuth converted into time.

7. The tangent of the angle θ, which Gauss designates by E, will also serve to calculate the angle at the star in the triangle formed by the pole, the zenith, and the star. This angle, formed

by the vertical and declination circles, which is called the *Parallactic angle*, is very often required. If we use the Auxiliary Tables mentioned before, in which the quantity E is tabulated, we obtain this angle, which shall be designated by p, by means of the convenient formula,

$$\tan p = \frac{E}{\cos (B + \delta)},$$

as is immediately seen if we apply to the right-angled triangle SGF, (fig. 1) the fifth of the formulæ (12) in No. 8 of the Introduction. If, on the contrary, we have not the Auxiliary Tables, we obtain by means of the formulæ of spherical trigonometry from the triangle SPZ,

$$\cos h \sin p = \cos \phi \sin t,$$
$$\cos h \cos p = \cos \delta \sin \phi - \sin \delta \cos \phi \cos t,$$

or if we put

$$\cos \phi \cos t = n \sin N,$$
$$\sin \phi = n \cos N,$$

for more convenient logarithmic computation,

$$\cos h \sin p = \cos \phi \sin t,$$
$$\cos h \cos p = n \cos (\delta + N).$$

The parallactic angle is used, amongst other cases, when we wish to determine the effect which a small change in the azimuth and altitude has upon the hour-angle and declination. We obtain namely, if we apply the first and third of formulæ (13), in No. 9 of the Introduction to the triangle between the pole, the zenith, and the star,

$$d\delta = \cos p \,.\, dh + \cos t \,.\, d\phi + \cos h \sin p \,.\, dA,$$

$$\cos \delta \,.\, dt = - \sin p \,.\, dh + \sin t \sin \delta \,.\, d\phi + \cos h \cos p \,.\, dA,$$

and similarly

$$dh = \cos p \,.\, d\delta - \cos A \,.\, d\phi - \cos \delta \sin p \,.\, dt,$$

$$\cos h \,.\, dA = \sin p . \, d\delta - \sin A \sin h . \, d\phi + \cos \delta \cos p . \, dt.$$

8. In order to transform the co-ordinates of right ascension and declination into co-ordinates of longitude and latitude, we

have only to turn the axis of z''* in the plane of $y''z''$ in the positive directions of y'' and z'' through the angle ϵ, which is equal to the obliquity of the ecliptic. Then we obtain from the formulæ (2) in No. 1 of the Introduction, since the axes of x'' and x''' coincide in the two systems:

$$\cos\beta\cos\lambda = \cos\delta\cos\alpha,$$

$$\cos\beta\sin\lambda = \cos\delta\sin\alpha\cos\epsilon + \sin\delta\sin\epsilon,$$

$$\sin\beta = -\cos\delta\sin\alpha\sin\epsilon + \sin\delta\cos\epsilon.$$

These formulæ we can also deduce in a different way by consideration of the triangle between the pole of the equator, the pole of the ecliptic, and the star, in which the three sides are $90^0 - \delta$, $90^0 - \beta$ and ϵ, and the angles respectively opposite to them are $90^0 - \lambda$, $90^0 + \alpha$, and the angle at the star.

In order to adapt the above formulæ for convenient logarithmic computation, we introduce the auxiliary quantities

$$\left.\begin{aligned} M\sin N &= \sin\delta \\ M\cos N &= \cos\delta\sin\alpha \end{aligned}\right\} \cdots\cdots\cdots\cdots (a_1),$$

by which means the three equations are changed into the following,

$$\cos\beta\cos\lambda = \cos\delta\cos\alpha,$$

$$\cos\beta\sin\lambda = M\cos(N-\epsilon),$$

$$\sin\beta = M\sin(N-\epsilon),$$

or, if we desire to express all the quantities by means of tangents, and substitute for M its value

$$\frac{\cos\delta\sin\alpha}{\cos N},$$

they are changed into the following:

$$\left.\begin{aligned} \tan N &= \frac{\tan\delta}{\sin\alpha} \\ \tan\lambda &= \frac{\cos(N-\epsilon)}{\cos N}\tan\alpha \\ \tan\beta &= \tan(N-\epsilon)\sin\lambda \end{aligned}\right\} \cdots\cdots\cdots\cdots (b_1).$$

* See No. 3 of this Section.

The primary formulæ give α and δ without any ambiguity; but, if the formulæ (b_1) are used for computation, it may be doubtful in which quadrant the angle λ is to be taken. But it follows from the equation

$$\cos\beta\cos\lambda = \cos\delta\cos\alpha,$$

that we must take the angle λ always in that quadrant which at once satisfies the sign of $\tan\lambda$, and besides fulfils the condition that $\cos\alpha$ and $\cos\lambda$ must have the same sign.

As a verification of the computation we can besides employ the equation

$$\frac{\cos(N-\epsilon)}{\cos N} = \frac{\cos\beta\sin\lambda}{\cos\delta\sin\alpha} \quad\ldots\ldots\ldots\ldots\ldots\ldots\quad (c_1),$$

which arises from the division of the equation

$$\cos\beta\sin\lambda = M\cos(N-\epsilon)$$

by

$$\cos\delta\sin\alpha = M\cos N.$$

The geometrical signification of the auxiliary quantities is easily discoverable. N is the angle which the great circle connecting the first point of Aries with the star forms with the equator, and M the sine of the arc of the great circle.

Example.

Let $\alpha = 6^\circ.33'.29'',30 \quad \delta = -16^\circ.22'.35'',45$

$\epsilon = 22^\circ.27'.31'',72.$

Then the computation of the formulæ (b_1) and (c_1) gives

$\cos\delta$	9·9820131	$\tan\alpha$	9·0605604
$\tan\delta$	−9·4681562	$\frac{\cos(N-\epsilon)}{\cos N}$	−9·0292017
$\sin\alpha$	9·0577093	$\lambda =$	$359^\circ.17'.43'',91$
$N =$	$-68^\circ.45'.41'',88$		
$\epsilon =$	$+23^\circ.27'.31'',72$	$\tan(N-\epsilon)$	1·4114653
$N-\epsilon =$	$-92^\circ.13'.13'',60$	$\sin\lambda$	−8·0897293
$\cos(N-\epsilon)$	−8·5882086	$\beta =$	$-17^\circ.35'.37'',53$
$\cos N$	9·5590069	$\cos\beta =$	9·9791948

$$\begin{aligned} \cos\beta\sin\lambda &= -\,8{\cdot}0689241 \\ \cos\delta\sin\alpha &= 9{\cdot}0397224 \\ \hline & -\,9{\cdot}0292017 \end{aligned}$$

Applying to the triangle between the star, the pole of the equator, and the pole of the ecliptic, the Gaussian formulæ, we obtain, indicating the angle at the star by $90 - E$*,

$$\sin\left(45^0 - \frac{\beta}{2}\right)\sin\frac{E-\lambda}{2} = \cos\left(45^0 + \frac{\alpha}{2}\right)\sin\left(45^0 - \frac{\epsilon+\delta}{2}\right),$$

$$\sin\left(45^0 - \frac{\beta}{2}\right)\cos\frac{E-\lambda}{2} = \sin\left(45^0 + \frac{\alpha}{2}\right)\cos\left(45^0 - \frac{\epsilon-\delta}{2}\right),$$

$$\cos\left(45^0 - \frac{\beta}{2}\right)\sin\frac{E+\lambda}{2} = \sin\left(45^0 + \frac{\alpha}{2}\right)\sin\left(45^0 - \frac{\epsilon-\delta}{2}\right),$$

$$\cos\left(45^0 - \frac{\beta}{2}\right)\cos\frac{E+\lambda}{2} = \cos\left(45^0 + \frac{\alpha}{2}\right)\cos\left(45^0 - \frac{\epsilon+\delta}{2}\right),$$

formulæ, which are particularly useful, if we desire to determine at the same time the quantities λ and β as well as the angle $90^0 - E$.

NOTE. Encke has in the *Jahrbuch* for 1831, given tables which, for an approximate computation of the longitude and latitude from the right ascension and declination, are extremely useful. They depend upon the second of the transformations of the three fundamental equations given in No. 6 of the Introduction, similarly to the tables specified in No. 6 of this section.

9. For the converse case, when it is required to transform the co-ordinates of a star referred to the ecliptic into co-ordinates referred to the equator, the formulæ are exactly similar. We obtain then by means of formulæ (1) for the transformation of co-ordinates or by means of the spherical triangle previously treated of,

$$\begin{aligned} \cos\delta\cos\alpha &= \cos\beta\cos\lambda, \\ \cos\delta\sin\alpha &= \cos\beta\sin\lambda\cos\epsilon - \sin\beta\sin\epsilon, \\ \sin\delta &= \cos\beta\sin\lambda\sin\epsilon + \sin\beta\cos\epsilon. \end{aligned}$$

* Gauss, *Theoria motus*, page 64.

We obtain also the same equations by replacing, in the three fundamental equations in No. 8, β and λ by δ and α, and taking the angle ϵ negatively. By this means we find from the formulæ (b_1),

$$\tan N = \frac{\tan\beta}{\sin\lambda}$$

$$\tan\alpha = \frac{\cos(N+\epsilon)}{\cos N}\tan\lambda$$

$$\tan\delta = \tan(N+\epsilon)\sin\alpha,$$

and, from (c_1), the check equation

$$\frac{\cos(N+\epsilon)}{\cos N} = \frac{\cos\delta\sin\alpha}{\cos\beta\sin\lambda},$$

where now N signifies the angle which the great circle connecting the star with the first point of Aries makes with the ecliptic.

The Gaussian equations give finally for this case:

$$\sin\left(45^0-\frac{1}{2}\delta\right)\sin\frac{1}{2}(E+\alpha)=\sin\left(45^0+\frac{1}{2}\lambda\right)\sin\left\{45^0-\frac{1}{2}(\epsilon+\beta)\right\},$$

$$\sin\left(45^0-\frac{1}{2}\delta\right)\cos\frac{1}{2}(E+\alpha)=\cos\left(45^0+\frac{1}{2}\lambda\right)\cos\left\{45^0-\frac{1}{2}(\epsilon-\beta)\right\},$$

$$\cos\left(45^0-\frac{1}{2}\delta\right)\sin\frac{1}{2}(E-\alpha)=\cos\left(45^0+\frac{1}{2}\lambda\right)\sin\left\{45^0-\frac{1}{2}(\epsilon-\beta)\right\},$$

$$\cos\left(45^0-\frac{1}{2}\delta\right)\cos\frac{1}{2}(E-\alpha)=\sin\left(45^0+\frac{1}{2}\lambda\right)\cos\left\{45^0-\frac{1}{2}(\epsilon+\beta)\right\}.$$

It is not necessary to introduce an example for this case, since the formulæ are precisely similar to the former.

NOTE. For the sun, which always moves in the ecliptic, the expressions become simpler. Denoting namely the sun's longitude by L, its right ascension and declination by A and D, we obtain,

$$\tan A = \tan L\cos\epsilon,$$
$$\sin D = \sin L\sin\epsilon,$$

or in addition,

$$\tan D = \tan\epsilon\sin A.$$

10. The angle at the star in the triangle formed by great circles joining the pole of the equator, the pole of the ecliptic, and the star, which is contained between the circle of declination and the circle of latitude, is found immediately in terms of λ and β or of α and δ, by employing the Gaussian formulæ for the computation of these quantities, since, if this angle be denoted by η, $\eta = 90^0 - E$. But if this angle is required, without the necessity of computing by the Gaussian formulæ, it is found by means of the equations:

$$\cos\beta \sin\eta = \cos\alpha \sin\epsilon,$$
$$\cos\beta \cos\eta = \cos\epsilon \cos\delta + \sin\epsilon \sin\delta \sin\alpha,$$

or,

$$\cos\delta \sin\eta = \cos\lambda \sin\epsilon,$$
$$\cos\delta \cos\eta = \cos\epsilon \cos\beta - \sin\epsilon \sin\beta \sin\lambda;$$

or, if we put

$$\cos\epsilon = m \cos M,$$
$$\sin\epsilon \sin\alpha = m \sin M,$$

or,

$$\cos\epsilon = n \cos N,$$
$$\sin\epsilon \sin\lambda = n \sin N,$$

by means of the equations

$$\cos\beta \sin\eta = \cos\alpha \sin\epsilon,$$
$$\cos\beta \cos\eta = m \cos (M - \delta),$$

or,

$$\cos\beta \sin\eta = \cos\lambda \sin\epsilon,$$
$$\cos\beta \cos\eta = n \cos (N + \beta).$$

This angle again is needed, when it is required to investigate the effect which small changes in the quantities λ, β, and ϵ will produce on α and δ, and *vice versâ*. We obtain, namely, by employing for the triangle under consideration the first and third of the formulæ (13) in No. 9 of the Introduction,

$$d\beta = \cos\eta\, d\delta - \cos\delta \sin\eta\, d\alpha - \sin\lambda\, d\epsilon,$$
$$\cos\beta\, d\lambda = \sin\eta\, d\delta + \cos\delta \cos\eta\, d\alpha + \cos\lambda \sin\beta\, d\epsilon,$$

and for the converse case,

$$d\delta = \cos\eta\, d\beta + \cos\beta \sin\eta\, d\lambda + \sin\alpha\, d\epsilon,$$
$$\cos\delta\, d\alpha = - \sin\eta\, d\beta + \cos\beta \cos\eta\, d\lambda - \cos\alpha \sin\delta\, d\epsilon.$$

11. For the sake of completeness we will in addition give the formulæ for the transformation of the first system of co-ordinates into the fourth; although these are never employed.

First with reference to the plane of the horizon:

$$x = \cos A \cos h,$$
$$y = \sin A \cos h,$$
$$z = \sin h.$$

If the axis of x be turned towards the positive direction of the axis of z in the plane of xz through an angle $90^0 - \phi$, we obtain the new co-ordinates

$$x' = x \sin\phi + z \cos\phi,$$
$$y' = y,$$
$$z' = z \sin\phi - x \cos\phi.$$

If now the axis of x' be turned in the plane of $x'y'$, which is the plane of the equator, through an angle Θ, so that the axis of x'' is in the direction of the first point of Aries, we obtain, considering that the positive axis of y'' is in the direction of the ninetieth degree of right ascension, and that the hour-angle and the right ascension are reckoned in opposite directions,

$$x'' = x' \cos\Theta + y' \sin\Theta,$$
$$-y'' = y' \cos\Theta - x' \sin\Theta,$$
$$z'' = z'.$$

Turn lastly the axis of y'' in the plane of $y''z''$ towards the positive direction of the axis of z'' through an angle ϵ, and we then obtain

$$x''' = x'',$$
$$y''' = y'' \cos\epsilon + z'' \sin\epsilon,$$
$$z''' = -y'' \sin\epsilon + z'' \cos\epsilon,$$

and, since we have besides,

$$x''' = \cos\beta \cos\lambda,$$
$$y''' = \cos\beta \sin\lambda,$$
$$z''' = \sin\beta,$$

we shall be able, through elimination of x', y', z', and x'', y'', z'', to express λ and β immediately in terms of A, h, ϕ, Θ, and ϵ.

III. *Particular Phenomena of the Daily Motion.*

12. In No. 6 was found the equation

$$\sin h = \sin\phi \sin\delta + \cos\phi \cos\delta \cos t.$$

When the object is in the horizon, $h=0$, and hence we have

$$0 = \sin\phi \sin\delta + \cos\phi \cos\delta \cos t_0,$$

or

$$\cos t_0 = -\tan\phi \tan\delta \; \ldots\ldots\ldots\ldots\ldots\ldots \; (a).$$

By means of this formula is found for a given latitude ϕ the hour-angle at rising and setting of a star whose declination is δ. The value of this hour-angle taken absolutely is termed the *semi-diurnal arc* of the star. If the sidereal time be known at which the star passes the meridian, or its right-ascension, the sidereal time of its rising or setting can be computed by subtracting from the right ascension or adding to it the absolute value of t_0. It follows moreover from this, that, on the equator only, where $\phi = 0$ and the semi-diurnal arc for all stars is equal to 90°, all stars which rise at the same time set also at the same time.

EXAMPLE. It is required to calculate at what time the star Arcturus rises and sets at Berlin. For the beginning of the year 1848

$$\alpha = 14^h.\,8^m,7, \quad \delta = +\,19^0.\,58',5\,;$$

in addition

$$\phi = 52^0.\,30'.\,16''.$$

We have therefore

$$\begin{array}{ll} \tan\delta & 9{\cdot}56048 \\ \tan\phi & 0{\cdot}11509 \\ \hline \end{array}$$

$$t_0 = 118^0.\,16',8$$
$$= 7^h.\,53^m.\,7^s.$$

Arcturus therefore rises at $6^h.\,15^m,6$ sidereal time and sets at $22^h.\,1^m,8$.

If δ be positive, and the star be therefore north of the equator, $\cos t_0$ will, for places in north latitudes, be negative; t_0 is therefore greater than 90°, and the star remains a longer time above the horizon than below it. For stars with south declination, on the contrary, t_0 will be less than 90°, and, for places in the northern hemisphere, these remain for a shorter time above the horizon than below it.

In the southern hemisphere, where ϕ has negative values, the converse holds good, since there the diurnal arc for southern stars is greater than 12 hours. If $\phi = 0$, t_0 will, for every value of δ, be equal to 90°; and therefore at the equator all stars remain for an equal time above and below the horizon. If $\delta = 0$, t_0 will, for every value of ϕ, be equal to 90°. Thus, for equatoreal stars, the time during which they are above the horizon is, for all places on the earth, equal to the time during which they are beneath the horizon.

Thus, if the sun be north of the equator, then, for the northern hemisphere of the earth, the days are longer than the nights, and conversely, when he is south of the equator. If however the sun be in the equator, then, for all places of the earth, day and night are equal. On the equator this is always the case. The value of t_0 will besides be possible only as long as $\tan\phi \tan\delta$ is less than 1. Thus, for a star to set at a place in latitude ϕ, it is necessary that $\tan\delta$ be less than $\cot\phi$, or δ less than $90^\circ - \phi$. If $\delta = 90^\circ - \phi$, then is $t_0 = 180^\circ$, and the star only touches the horizon at its lower culmination. If δ be greater than $90^\circ - \phi$, the star never sets; if, on the contrary, the south declination be greater than $90^\circ - \phi$, the star never ascends above the horizon.

Since the declinations of the sun lie always between $-\epsilon$ and $+\epsilon$, it follows that all places of the earth for which the sun for only one day of the year does not rise or set, must have a north or south latitude equal to $90^\circ - \epsilon$ or to $66\frac{1}{2}^\circ$. These places lie in the two polar circles. Places lying still nearer to the pole of the earth have the sun in summer a still longer time without interruption above the horizon, and in winter beneath it, the nearer they lie to the pole.

NOTE. The equation for the hour-angle at rising and setting may still be brought into another form. By subtracting equation (α) from unity and also adding it to unity, we get, by division of the two new equations,

$$\tan^2 \frac{1}{2} t_0 = \frac{\cos(\phi - \delta)}{\cos(\phi + \delta)}.$$

This equation also shews that t_0 is only possible as long as

$\cos(\phi - \delta)$ and $\cos(\phi + \delta)$ are positive, and that therefore those stars only whose south or north declinations are less than $90^0 - \phi$, can for this latitude rise or set.

13. To find the point of the horizon where a star rises or sets, we have only, in the equation given in No. 5,

$$\sin\delta = \sin\phi \sin h - \cos\phi \cos h \cos A$$

to make $h = 0$, whereby we obtain

$$\cos A_0 = -\frac{\sin\delta}{\cos\phi} \quad\ldots\ldots\ldots\ldots\ldots\ldots\ldots \quad (b).$$

The negative value of A_0 is the azimuth of the star at its rising, the positive value is the azimuth at its setting. The distance of the star from the true east or west point is termed the morning or evening *amplitude* of the star. Denoting this by A_1, we have

$$A_0 = 90^0 + A_1;$$

and thus

$$\sin A_1 = \frac{\sin\delta}{\cos\phi} \quad\ldots\ldots\ldots\ldots\ldots\ldots\ldots\ldots \quad (c),$$

where A_1 is positive when the point of rising or setting lies on the north of the east or west point, negative when it lies on the south.

The formula (*c*) for the morning and evening amplitude can again be also given in another shape, by writing

$$\frac{1 + \sin A_1}{1 - \sin A_1} = \frac{\sin\psi + \sin\delta}{\sin\psi - \sin\delta},$$

where $\psi = 90^0 - \phi$. Hence we obtain

$$\tan^2\left(45^0 - \frac{A_1}{2}\right) = \frac{\tan\frac{\psi - \delta}{2}}{\tan\frac{\psi + \delta}{2}}.$$

For Arcturus we accordingly obtain, with the preceding values of δ and ϕ,

$$A_1 = 34^0.\,8',3.$$

14. By putting in the equation

$$\sin h = \sin \phi \sin \delta + \cos \phi \cos \delta \cos t$$

$1 - 2 \sin^2 \frac{1}{2} t$ for $\cos t$, we obtain

$$\sin h = \cos (\phi - \delta) - 2 \cos \phi \cos \delta \sin^2 \frac{1}{2} t.$$

From this it is seen, in the first place, that, to equal values of t on both sides of the meridian, equal altitudes correspond. In addition, since the second term is always negative, h has, for $t = 0$, a maximum value, and this maximum, or the altitude of the star at its upper culmination, is given from the equation

$$\sin h = \cos (\phi - \delta) \quad \ldots\ldots\ldots\ldots\ldots\ldots (d).$$

For the lower culmination, or for $t = 180^\circ$, h will, on the contrary, be a minimum, as will be most easily seen by introducing $180^\circ + t'$ for t, where t' is reckoned from the northern part of the meridian. We shall then have, namely,

$$\sin h = \sin \phi \sin \delta - \cos \phi \cos \delta \cos t';$$

or, if again we put $1 - 2 \sin^2 \frac{1}{2} t$ for $\cos t'$,

$$\sin h = \cos \ (180^\circ - \phi - \delta) + 2 \cos \phi \cos \delta \sin^2 \frac{1}{2} t'.$$

Since both terms of the right side of the equation are now positive, $\sin h$, and consequently h, must, at the lower culmination of the star, be a minimum, and, for its value,

$$\sin h = \cos (180^\circ - \phi - \delta) \quad \ldots\ldots\ldots\ldots\ldots\ldots (e).$$

From the equation (d) it follows, that $90^\circ - h$, or the zenith-distance of the star at its upper culmination, is either $\phi - \delta$, or $\delta - \phi$. But, since the zenith-distance must always be positive, we must, so long as the star culminates on the south side of the zenith, that is so long as δ is less than ϕ, take for the zenith-distance $\phi - \delta$. If however the star culminates on the north side of the zenith, where δ must be greater than ϕ, we must take $\delta - \phi$ for the zenith-distance. For the zenith-distance at the lower culmination we obtain from equation (e)

$$z = 180^\circ - \phi - \delta.$$

To bring all the three cases under one algebraical formula, we take as the common expression for the zenith distance of a star at its passage over the meridian

$$z = \delta - \phi \ldots\ldots\ldots\ldots\ldots\ldots\ldots\ldots (f).$$

We must then consider south zenith distances as negative, and at the lower culmination take $180^0 - \delta$ instead of δ, or we must, in the last case, reckon δ from that point of the equator which cuts the visible meridian.

The declination of α Lyræ is $38^0.39'$, and thus, for the latitude of Berlin, $\delta - \phi = -13^0.51'$. The star α Lyræ passes thus at its upper culmination for Berlin south of the zenith at a distance of $13^0.51'$. In addition $180^0 - \phi - \delta$, or the zenith distance at the lower culmination, is equal to $88^0.51'$.*

15. The greatest altitude of a star takes place on the meridian, only when its declination during the time of its continuance about the horizon does not change. If, on the contrary, the declination is variable, the body attains its greatest altitude off the meridian. By differentiating the formula

$$\cos z = \sin \phi \sin \delta + \cos \phi \cos \delta \cos t,$$

regarding z, δ, and t as variable, we obtain

$$-\sin z dz = [\sin \phi \cos \delta - \cos \phi \sin \delta \cos t]\, d\delta,$$
$$-\cos \phi \cos \delta \sin t\, dt,$$

and hence, for the case when z is a maximum or $dz = 0$,

$$\sin t = \frac{d\delta}{dt}\, [\tan \phi - \tan \delta \cos t].$$

From this equation is found the hour-angle of the body at the time of its greatest altitude. $\frac{d\delta}{dt}$ is the ratio of the change of declination to the change of hour-angle, so that if, for example, dt represents a second of arc, $\frac{d\delta}{dt}$ is the change of declination in $\frac{1}{15}$ of a second of time. Since this ratio for all the heavenly

* [In the preceding remarks zenith-distances are reckoned positively towards the North, but it is more usual as well as more convenient to measure positively towards the South, as well as to measure azimuths from the South point of the horizon.—TRANSLATOR.]

bodies is small, we may replace sin t by the arc and put unity for cos t, and we then obtain for the hour-angle corresponding to the greatest altitude

$$*t = \frac{d\delta}{dt}\,[\tan\phi - \tan\delta] \times \frac{206265}{15} \quad \ldots\ldots\ldots\ldots (g),$$

where $\frac{d\delta}{dt}$ is the change of declination in a second of time, and t is expressed in seconds of time. This hour-angle t must then always be added algebraically to the time of culmination, to obtain the time of the greatest altitude.

If the body culminate south of the zenith, and if it be approaching the north pole, $\frac{d\delta}{dt}$ is positive, and therefore, when ϕ is positive, the greatest altitude takes place after the culmination; if, on the contrary, the declination be diminishing, the greatest altitude precedes the culmination. The converse takes place when the body culminates between the zenith and the pole†.

16. Differentiating the formula

$$\sin h = \sin\phi \sin\delta + \cos\phi \cos\delta \cos t$$

with respect to h and t, we obtain for the change of altitude of a body,

$$\cos h \frac{dh}{dt} = -\cos\phi \cos\delta \sin t,$$

or

$$\frac{dh}{dt} = -\cos\delta \sin p \quad \ldots\ldots\ldots\ldots (h),$$

where $\cos h \sin p = \cos\phi \sin t$, according to No. 7 of this section.

In many cases also the second differential coefficient is needed. It is

$$\frac{d^2h}{dt^2} = -\cos\delta \cos p \frac{dp}{dt}.$$

* $[206265 = \frac{1}{\sin 1''}$.—TRANSLATOR.]

† [This *may* take place for European latitudes, but even then but rarely, in the motions of comets.—TRANSLATOR.]

Differentiating now the equation

$$\sin \phi = \sin h \sin \delta + \cos h \cos \delta \cos p,$$

considering h and p as variable, we obtain

$$0 = [\cos h \sin \delta - \sin h \cos \delta \cos p] \frac{dh}{dt} - \cos h \cos \delta \sin p \frac{dp}{dt},$$

or,

$$\frac{dp}{dt} = -\frac{\cos \phi \cos A}{\cos h \cos \delta \sin p} \cdot \frac{dh}{dt}$$

$$= + \frac{\cos \phi \cos A}{\cos h}.$$

Substituting this expression for $\frac{dp}{dt}$ in the equation $\frac{d^2h}{dt^2}$, we obtain

$$\frac{d^2h}{dt^2} = -\frac{\cos \delta \cos \phi}{\sin z} \cos A \cos p \text{ .... } (i).$$

In the same manner we obtain

$$\frac{dz}{dt} = + \cos \delta \sin p,$$

$$\frac{d^2z}{dt^2} = + \frac{\cos \delta \cos \phi}{\sin z} \cos A \cos p \text{ .... } (k).$$

17. Since $\cos \delta \sin p = \cos \phi \sin A$, we have also

$$\frac{dh}{dt} = -\cos \phi \sin A.$$

We shall thus have $\frac{dh}{dt} = 0$, and h a maximum or minimum, when $A = 0$, and so the star is on the meridian, and the second differential coefficient shews that h is a maximum when $A = 0$, and a minimum when $A = 180^\circ$.

In addition $\frac{dh}{dt}$ will be a maximum, if $\sin A = \pm 1$, or $A = 90^\circ$ or 270°. The altitude of a star is thus shewn to change most rapidly at the instant when it passes across the vertical circle, of which the azimuth is 90° or 270°. This vertical circle is called the *prime vertical.*

To find the time of passage of the star across the prime vertical, as well as its altitude, we have only, in formula No. 6

of this section, to make $A = 90^\circ$, or to solve the right-angled triangle between the star, the zenith, and the pole, and we obtain

$$\left.\begin{aligned} \cos t &= \frac{\tan\delta}{\tan\phi} \\ \sin h &= \frac{\sin\delta}{\sin\phi} \end{aligned}\right\} \quad \ldots\ldots\ldots\ldots\ldots\ldots\ldots\ldots \quad (l).$$

If δ be greater than ϕ, cos t will be impossible, and therefore the star cannot transit the prime vertical, but will culminate between the zenith and the pole. If δ be negative, cos t will be negative, but, since in northern latitudes the hour-angles of southern stars are always less than 90° as long as they are above the horizon, they cannot transit the visible portion of the prime vertical *.

For Arcturus, with the latitude of Berlin, we obtain

$$t = 73^\circ.48',5 = 4^h.55^m.14^s$$

and

$$h = 25^\circ.30',2.$$

Thus, for Berlin, Arcturus transits the prime vertical before the culmination at $9^h.15^m.5^s$, and after it, at $19^h.3^m.9^s$ sidereal time. If the hour-angle be nearly equal to 0, t is found by means of the cosine and h by the sine or tangent very incorrectly. We obtain in this case, from the formulæ for cos t, in the same manner as before,

$$\tan^2\frac{1}{2}t = \frac{\sin(\phi - \delta)}{\sin(\phi + \delta)},$$

and, for the computation of the altitude, we must take the following formula,

$$\text{cotan}\, h = \tan t \cos\phi.$$

IV. *On the Daily Motion as a Measure of Time.*

SIDEREAL TIME, SOLAR TIME, MEAN TIME.

18. Since the daily revolution of the sphere of the heavens, or, more properly, the revolution of the earth on its axis, is perfectly uniform, it serves us as a measure of time, which we

* As may be seen also by the equation for sin h, which then gives a negative value for h.

can have an idea of in like manner as being uniform in its progress. The time occupied by the earth in one revolution on its axis, and consequently the time which elapses between two consecutive culminations of the same fixed star, is called a *sidereal day*. The commencement of this day is reckoned, or it is said to be 0^h of sidereal time, at the instant when the first point of Aries passes the meridian. In the same manner it is said to be 1^h, 2^h, 3^h, &c. sidereal, when the hour-angle of the first point of Aries is 1^h, 2^h, 3^h, &c., and therefore when that point of the equator culminates whose right ascension is 1^h, 2^h, 3^h, &c., or 15°, 30°, 45°, &c.

We shall see in the sequel, that the equinoctial points, that is, the points of intersection of the ecliptic and the equator, are not fixed points, but that they have a retrograde motion along the equator. This motion is composed of two others, of which one is proportional to the time, and is connected with the daily motion of the heavens, but the other is periodical. Owing to this latter motion, the hour-angle of the first point of Aries does not change with perfect uniformity, and therefore the sidereal time is not a perfectly uniform measure. This want of uniformity is however exceedingly small, since the period of nineteen years has only the two maxima -1^s and $+1^s$.

19. When the sun on the 21st of March is at the vernal equinox, he passes the meridian very nearly at 0^h sidereal. But the sun now moves forward in the ecliptic, and, since on the 23rd of September he is in the autumnal equinox, and thus has 12^h right ascension, he culminates on this day at 12^h sidereal. The time of culmination, and, in like manner, the time of rising and setting of the sun runs through therefore in the course of a year all the hours of the sidereal day, and, on account of this inconvenience, sidereal time is not employed in the civil affairs of life, but the sun himself is used for the purpose of measuring time. The hour-angle of the sun at any time is called *true solar time*, and the time which elapses between two consecutive culminations of the sun is called a *true solar day*. At any place it is 0^h of true time when the sun passes the meridian of that place.

This *true time* has however this inconvenience, that it does not progress uniformly, since the right ascension of the sun does not change uniformly. In the first place, namely, the sun does not move in the equator but in the ecliptic, and we obtain his right ascension α from his longitude according to the Note to No. 9, by means of the formula

$$\tan \alpha = \tan \lambda \cos \epsilon,$$

or, if for this purpose we employ formula (19) in No. 11 of the Introduction, by means of the series

$$\alpha = \lambda - \tan^2 \frac{1}{2} \epsilon \sin 2\lambda + \frac{1}{2} \tan^4 \frac{1}{2} \epsilon \sin 4\lambda - \&c.$$

From this it is seen that the right ascension of the sun increases irregularly, even when the longitude increases uniformly. But besides this the sun moves also irregularly in his orbit, and Theoretical Astronomy teaches that his longitude at any time t is represented by an expression of the form

$$\lambda = L + \mu t + \zeta,$$

where ζ is a periodical function depending upon the longitude of the sun. From both causes then the right ascension of the sun increases irregularly, and consequently also his hour-angle or the true solar time. Since now our clocks have an uniform motion, and so cannot give true solar time, true time cannot be used for the ordinary purposes of life; an uniform time therefore is used which is called Mean Solar Time.

20. Between two successive transits of the sun through the vernal equinox 366·24222 sidereal days elapse, and therefore any particular star will in this time, which is called the *tropical year*, as often complete its daily revolution on the sphere of the heavens, or as often pass over the meridian. But, since the sun by its proper motion in the ecliptic, has likewise in that time passed through the 24 hours of the equator, so will it during a tropical year pass exactly once less across the meridian than a fixed star, viz. 365·24222 times. The tropical year has been divided into the same number of equal days, which are called *Mean Days*, and each one of these includes 24 equal hours, so that the tropical year is, when expressed in mean time, equal to

$$365^{d}.\ 5^{h}.\ 48^{m}.\ 47^{s},8091.$$

Imagine then that a fictitious sun moves in the equator with uniform velocity so that the right ascension α for any time t is obtained by means of the expression

$$\alpha = L + \mu t,$$

where L is the mean longitude of the sun at the commencement of the time t, and

$$\mu = \frac{360^0}{365 \cdot 24222};$$

and therefore if t be expressed in mean days its value $= 59'.8'',33$, and the hour-angle of this mean sun will be the *mean* time. The mean day begins when the sidereal time is equal to the longitude of the mean sun, or when this fictitious mean sun is on the meridian. In astronomical calculations the hours are reckoned from 0^h to 24^h.

Now the mean sun is sometimes before the true sun and sometimes behind it, according to the sign of the term ζ and of the periodical term $-\tan^2\frac{\epsilon}{2}\sin 2\lambda + \ldots$, which is called the *Reduction to the Ecliptic.* This difference between mean and true solar time is called the *Equation of time*, and its algebraical sign is so taken that it shall be always additive algebraically to true time to obtain mean time. The equation of time is zero four times in the year, or the mean time is the same as the true time, namely, on the 14th of April, the 14th of June, the 31st of August, and the 23rd of December, or on the day following these. Between the 23rd of December and the 14th of April the equation of time reaches its maximum value of $14^m.34^s$ about the middle of February, and during this period true time is behind mean time. Between the 14th of April and the 14th of June it reaches a maximum of $3^m.54^s$ about the middle of May, and during this period true time is before mean time. Between the 14th of June and the 31st of August, a maximum $6^m.11^s$ is attained about the end of July, and during this period true time is again behind mean time. Lastly, between the 31st of August and the 23rd of December, the maximum $16^m.17^s$ is attained about the middle of November, and here true time is again before mean time. The equation of time is given in astronomical ephemerides, and it is given in Encke's *Jahrbuch*, for every true Berlin noon*. The duration of a true day reaches a maximum at the end of December, when it amounts to

$$24^h.\ 0^m.\ 30^s{,}0.$$

The minimum, which occurs at the end of September, on the contrary, attains the value of $23^h.\ 59^m.\ 39^s{,}0$.

There are now three separate kinds of time used in astronomy, and it is therefore necessary to obtain a knowledge of the rules for the mutual transformation of these times.

21. *Transformation of mean time into sidereal time and vice versâ.*

Since 365·24222 mean days are equal to 366·24222 sidereal days,

$$\text{one sidereal day} = \frac{365{\cdot}24222}{366{\cdot}24222}\text{ mean day}$$

$$= \text{a mean day} - 3^m.\ 55^s{,}909 \text{ mean time,}$$

$$\text{and a mean day} = \frac{366{\cdot}24222}{365{\cdot}24222}\text{ sidereal day}$$

$$= \text{a sidereal day} + 3^m.\ 56^s{,}555 \text{ sidereal time.}$$

Thus, θ being the sidereal time, M the mean time and θ_0 the sidereal time for $M = 0$, that is for the commencement of the mean day or for mean noon, we have

$$M = (\theta - \theta_0)\,.\,\frac{24^h - 3^m.\ 55^s{,}909}{24^h},$$

and

$$\theta = \theta_0 + M\,.\,\frac{24^h + 3^m.\ 56^s{,}555}{24^h}.$$

Thus, in order to transform sidereal time into mean time and vice versâ, it is necessary to know the sidereal time at mean

* The calculation of the equation of time is deduced from the Solar Tables, from which we can find the mean and true longitudes and likewise the mean and true right ascensions of the sun for any given time. The best Solar Tables are those of Carlini, corrected by Bessel, in the *Effemeridi Astronomiche di Milano per l'anno* 1844. [Since the publication of *Brunnow's Astronomy*, Hansen's and Olufsen's *Solar Tables* have been published, as also Le Verrier's. Both are of great excellency, and far superior to Carlini's.—TRANSLATOR.]

noon, or the right ascension of the mean sun at the beginning of the mean day, and since this increases daily at the rate of $3^m.\,56^s{,}555347$, it needs be given only for some particular epoch. In astronomical ephemerides this quantity is given for convenience for every mean noon. For further simplification of the calculation there are other tables, which give the values of

$$\frac{24^h - 3^m.\,55^s{,}909}{24^h}\,t,$$

and

$$\frac{24^h + 3^m.\,56^s{,}555}{24^h}\,t,$$

for different values of the time t.

Such tables are likewise found in the astronomical ephemerides and in all collections of astronomical tables.

EXAMPLE. To transform 1849, June 9, $14^h.\,16^m.\,36^s{,}35$ sidereal time for Berlin into mean time.

The sidereal time at mean noon for this day in Encke's *Jahrbuch*, is

$$5^h.\,10^m.\,48^s{,}30,$$

consequently $9^h.\,5^m.\,48^s{,}05$ sidereal have elapsed between mean noon and the given time, and this, according to the auxiliary tables, or by actually performing the multiplication by the factor

$$\frac{24^h - 3^m.\,55^s{,}909}{24^h}, \text{ produces } 9^h.\,4^m.\,18^s{,}63 \text{ mean time.}$$

If the mean time be given it must be transformed by the auxiliary tables into sidereal time, and this must be added to the sidereal time at mean noon for the purpose of finding the sidereal time corresponding to the mean time.

22. *Transformation of true time into mean time and vice versâ.*

In order to transform true time into mean time we have only to take out from the ephemerides the equation of time for the given true time and to apply this algebraically to the given time.

According to the Berlin *Jahrbuch,* we have for the equation of time at true noon

		1st Diff.	2nd Diff.
1849 June 8	$-1^m.\ 20^s,73$		
		$+11^s,36$	
9	$-1\ .\ \ 9,37$		$+0^s,27$
		$+11,63$	
10	$-0\ .\ 57,74$		

If the true time for June 9 be $9^h.\ 5^m.\ 23^s,60$, we shall find the equation of time to be $-1^m.\ 4^s,98$; and therefore the mean time will be $9^h.\ 4^m.\ 18^s,62$.

In order to transform mean time into true the same equation of time will serve. But since this is given in the ephemerides for true time, it is necessary to have a knowledge of the true time, in order to be able to interpolate the equation of time. But owing to the small daily change of this quantity it will be sufficient, when we are transforming the given mean time into true, to add to the given time an equation which corresponds only approximately to the given time. The equation of time is then interpolated with this approximate true time. If, for example, the mean time $9^h.\ 4^m.\ 18^s,62$ be given, we take for the equation of time -1^m. With the true time $9^h.\ 5^m.\ 18^s,6$ the equation of time will be found to be $-1^m.\ 4^s,98$, and consequently the true time is $9^h.\ 5^m.\ 23^s,60$.

23. *Transformation of true time into sidereal time and vice versâ.*

Since true time is nothing else than the hour-angle of the sun, his right ascension A needs only be known in order to obtain the sidereal time from the equation

$$\theta = W + A,$$

where W represents the true time.

According to Encke's *Jahrbuch,* we have the following right ascensions of the sun for true noon at Berlin:

		1st Diff.	2nd Diff.
1849 June 8	$5^h.\ 5^m.\ 30^s,79$		
		$+4^m.\ 7^s,96$	
9	$9\ .\ 38,75$		$+0^s,27$
		$4\ .\ 8,23$	
10	$13\ .\ 46,98$		

Should now the true time $9^h.\ 5^m.\ 23^s,60$ for June 9 be required to be transformed into sidereal time, we have for this time the right ascension of the sun equal to $5^h. 11^m. 12^s,75$, and consequently the sidereal time equal to $14^h. 16^m. 36^s,35$.

In order to transform sidereal time into true time, an approximate knowledge of the true time is required for the interpolation of the right ascension of the sun. But if from the given sidereal time be subtracted the right ascension of the sun which answers to the commencement of the day, the number of sidereal hours is obtained which have elapsed since that time. These sidereal hours must be transformed into true time. But it is sufficient to transform them into mean time and to interpolate the right ascension of the sun for this mean time. If then we subtract this from the given sidereal time, we obtain the true time.

The right ascension of the sun at the beginning of June 9 is equal to $5^h. 9^m. 38^s,75$, consequently between this and $14^h. 16^m. 36^s,35$ sidereal time, $9^h. 6^m. 57^s,60$ have elapsed, or $9^h. 5^m. 28^s,00$ of mean time. By interpolating the right ascension of the sun for this time, we obtain again $5^h. 11^m. 12^s,75$, and consequently the true time equal to $9^h. 5^m. 23^s,60$.

These transformations can moreover be performed with equal facility when the mean time is required from the sidereal, and from that the true time by means of the equation of time.

SECOND SECTION.

CORRECTIONS OF OBSERVATIONS, WHICH ARE DEPENDENT ON THE POSITION OF THE OBSERVER ON THE SURFACE OF THE EARTH AND ON THE PROPERTIES OF LIGHT.

THE Astronomical Tables and Ephemerides always give the places of the heavenly bodies as they would appear from the centre of the earth. For bodies infinitely distant, this place is the same as that which would be observed from any point upon the surface of the earth. But, if the distance of the body bear an appreciable ratio to the radius of the earth, the body will not appear to be in the same position when observed at the centre and at a point on the surface of the earth. Should it therefore be required to compare the observed place of such a body with the tables, it is necessary to devise a method by which the place as seen from the centre of the earth may be calculated from the observed place. Should it be desired, on the other hand, to calculate other quantities from the observed places of such a body referred to the horizon of the observer, for example, in connection with his known position with reference to the equator, the apparent place must then be employed as it appears when seen from the place of observation, and consequently the places as seen from the centre of the earth, which are given in the Ephemerides, must be transformed into apparent places.

The angle at the star which is included between the two lines drawn respectively from the centre of the earth and the place of observation on the surface to the star, is called the *Parallax.* Consequently a method is required by which the parallax of a star for any time and place upon the surface of the earth can be calculated.

Our earth is in addition surrounded, to some considerable altitude, by an atmosphere which possesses the property of refracting light. The heavenly bodies therefore are not seen in their true places, but in the direction which the ray of light, after being refracted by the atmosphere, has at the instant when it meets the eye of the observer.

The difference between this direction and that in which the heavenly body would be seen, supposing no atmosphere to exist, is called the *Refraction*. In order therefore to obtain the true place of a heavenly body from the observed place, a method is required by which the refraction may be determined for every point of the heavens and for every condition of the atmosphere.

Had the earth no proper motion, or were the velocity of light infinitely greater than the velocity of the earth, this motion would have no influence upon the apparent places of the heavenly bodies. But, since the velocity of light bears an appreciable ratio to the velocity of the earth, so an observer upon the earth sees all the heavenly bodies in advance of their true places by a small angle which is dependent upon this ratio, and towards that direction in which the earth is moving. This minute angle by which the places of the heavenly bodies appear to be altered by the motion of the earth and of light, is called the *Aberration*. In order therefore to obtain the true places of the heavenly bodies from observation means must be devised for freeing the observed apparent places from this effect of aberration.

I. Parallax.

1. Our earth is not a perfect sphere, but an oblate spheroid, that is to say, such as is formed by the revolution of an ellipse about its minor axis. Denoting by a the semi-major, by b the semi-minor axis, and by α the ellipticity in parts of the semi-major axis, we have

$$\alpha = \frac{a-b}{a} = 1 - \frac{b}{a}.$$

If moreover ϵ be the excentricity of the generating ellipse, that is that ellipse which is formed by the intersection of the surface of the spheroid and a plane passing through the minor axis, we have, if this excentricity be expressed likewise in parts of the semi-axis major,

$$\epsilon^2 = 1 - \frac{b^2}{a^2},$$

and therefore $$\frac{b}{a} = \sqrt{(1 - \epsilon^2)}.$$

Hence $$\alpha = 1 - \sqrt{(1 - \epsilon^2)},$$

and $$\epsilon = \sqrt{(2\alpha - \alpha^2)}.$$

The ratio $\frac{b}{a}$ is, according to Bessel's investigations, in the case of the earth,

$$\frac{298 \cdot 1528}{299 \cdot 1528},$$

or $$\alpha = \frac{1}{299 \cdot 1528},$$

and, expressed in toises,

$$a = 3272077 \cdot 14 \qquad \log a = 6 \cdot 5148235$$
$$b = 3261139 \cdot 33 \qquad \log b = 6 \cdot 5133693.$$

But in astronomy it is not the toise but the semi-axis major of the orbit of the earth which is taken as the unit. Denoting then by π the angle under which the equatoreal radius of the earth, or the semi-axis major of the spheroid, appears when viewed from the sun, and by R the semi-axis major of the earth's orbit, or the mean distance of the earth from the sun, we have

$$a = R \sin \pi = R\pi \sin 1'',$$

or $$a = \frac{R\pi}{206265}.$$

The angle π, or the equatoreal horizontal parallax of the sun is, according to Encke,

$$8'',57116.$$

This is the angle under which the radius of the earth's equator is seen from the sun, when the sun for places on the equator is in the horizon.

2. In order then to be able to calculate the parallax of a body for any place upon the surface, we must be able by means of co-ordinates to refer to the centre any point upon the spheroid. Take then as the first co-ordinate the sidereal time, that is the angle which a plane* passing through the place of observation and the semi-axis minor makes with a plane passing through the semi-axis minor and the first point of Aries. If then OAC (fig. 2) be the plane passing through the place of observation A and the semi-axis minor, we must, for the determination of the position of the place, know in addition the distance $AO = \rho$ from the centre of the earth, and the angle AOC, which is called the *corrected latitude.*

But these quantities can always be calculated from the astronomical latitude ANC, (which is in fact the angle which the horizon at A makes with the axis of the earth, or which the normal AN to the surface at A makes with the equator,) and the two axes of the earth's spheroid.

Let x and y be the co-ordinates of the point A referred to the centre O, considering OC as the axis of abscissæ and OB as the axis of ordinates; we have then, since A is a point in an ellipse of which a and b are respectively the semi-axes major and minor, the equation

$$a^2y^2 + b^2x^2 = a^2b^2.$$

Since now, if we denote by ϕ' the corrected latitude

$$\tan \phi' = \frac{y}{x},$$

and also

$$\tan \phi = -\frac{dx}{dy},$$

where the latitude ϕ is the angle which the normal at A makes with the axis of the abscissæ, we obtain, since the differential equation to the ellipse gives

$$\frac{y}{x} = -\frac{b^2}{a^2}\frac{dx}{dy},$$

* Since this plane passes through the poles of the earth and the zenith of the place of observation, it is the plane of the meridian.

the following relation between the quantities ϕ and ϕ',

$$\tan\phi' = \frac{b^2}{a^2}\,.\tan\phi \quad\ldots\ldots\ldots\ldots\ldots\ldots\ldots\ldots\ldots (a).$$

In order to calculate ρ, we have

$$\rho = \sqrt{(x^2+y^2)} = \frac{x}{\cos\phi'}.$$

Since now from the equation to the ellipse,

$$x = \frac{a}{\sqrt{\left(1+\frac{a^2}{b^2}\tan^2\phi'\right)}} = \frac{a}{\sqrt{(1+\tan\phi\tan\phi')}},$$

we thence obtain

$$\rho = \frac{a\sec\phi'}{\sqrt{(1+\tan\phi\tan\phi')}} = a\sqrt{\left\{\frac{\cos\phi}{\cos\phi'\cos(\phi'-\phi)}\right\}} \quad\ldots\ldots (b).$$

By means of these two formulæ we can calculate the corrected latitude ϕ' and the radius ρ for any place on the surface of the earth of which the latitude ϕ is known.

For the co-ordinates x and y we obtain the following formulæ, which also will be required in the sequel:

$$x = \frac{a\cos\phi}{\sqrt{\{\cos^2\phi + (1-\epsilon^2)\sin^2\phi\}}}$$

$$= \frac{a\cos\phi}{\sqrt{(1-\epsilon^2\sin^2\phi)}} \quad\ldots\ldots\ldots\ldots\ldots\ldots\ldots\ldots\ldots (c),$$

and

$$y = x\tan\phi' = x\frac{b^2}{a^2}\tan\phi = x(1-\epsilon^2)\tan\phi$$

$$= \frac{a(1-\epsilon^2)\sin\phi}{\sqrt{(1-\epsilon^2\sin^2\phi)}} \quad\ldots\ldots\ldots\ldots\ldots\ldots\ldots\ldots\ldots (d).$$

From the formula (a) we can express ϕ' in a series which proceeds according to the sines of the multiples of ϕ. We obtain, namely from the formulæ (18) in No. 11 of the Introduction,

$$\phi' = \phi - \frac{a^2-b^2}{a^2+b^2}\sin 2\phi + \frac{1}{2}\left(\frac{a^2-b^2}{a^2+b^2}\right)^2\sin 4\phi - \&c. \ldots\ldots (A)$$

or, putting $$\frac{a-b}{a+b}=n,$$

$$\phi'=\phi-\frac{2n}{1+n^2}\sin 2\phi+\frac{1}{2}\left(\frac{2n}{1+n^2}\right)^2\sin 4\phi-\&c. \;\ldots\ldots (B).$$

The angle $\phi-\phi'$ is called the *Angle of the Vertical.*

If the numerical values of the coefficient for the ellipticity of the earth given above be calculated and multiplied by 206265 $\left(\text{or } \frac{1}{\sin 1''}\right)$ in order to obtain the result in seconds, we arrive at

$$\phi'=\phi-11'.30'',65\sin 2\phi+1'',16\sin 4\phi-\&c. \;\ldots\ldots (C),$$

from which, for example, we find for the latitude of Berlin

$$52^0.30'.16'',0,$$

$$\phi'=52^0.19'.8'',3.$$

Although ρ itself cannot be expressed in so symmetrical a series as ϕ', a similar one can be obtained for $\log\rho$*.

The formula (b) gives for example:

$$\rho^2=\frac{a^2}{\cos^2\phi'\left(1+\frac{b^2}{a^2}\tan^2\phi\right)},$$

and, putting for $\cos^2\phi'$ its value $\frac{a^4}{a^4+b^4\tan^2\phi}$,

we obtain

$$\rho^2=\frac{a^4\cos^2\phi+b^4\sin^2\phi}{a^2\cos^2\phi+b^2\sin^2\phi}$$

$$=\frac{a^4+b^4+(a^4-b^4)\cos 2\phi}{a^2+b^2+(a^2-b^2)\cos 2\phi}$$

$$=\frac{(a^2+b^2)^2+(a^2-b^2)^2+2(a^2+b^2)(a^2-b^2)\cos 2\phi}{(a+b)^2+(a-b)^2+2(a+b)(a-b)\cos 2\phi}$$

and consequently,

$$\therefore\ \rho=\frac{(a^2+b^2)\left\{1+\left(\frac{a^2-b^2}{a^2+b^2}\right)^2+2\frac{a^2-b^2}{a^2+b^2}\cos 2\phi\right\}^{\frac{1}{2}}}{(a+b)\left\{1+\left(\frac{a-b}{a+b}\right)^2+2\frac{a-b}{a+b}\cos 2\phi\right\}^{\frac{1}{2}}}.$$

* Encke, *Jahrbuch for* 1852, p. 326, where tables are likewise given from which $\log\rho$ can be found for any value of ϕ'.

If the formula be written logarithmically and the logarithms of the square roots be developed according to the formula (17) in No. 11 of the Introduction, in series which proceed according to the cosines of the multiples of 2ϕ, we obtain:

$$\log_e \rho = \log_e . \frac{a^2+b^2}{a+b} + \left(\frac{a^2-b^2}{a^2+b^2} - \frac{a-b}{a+b}\right)\cos 2\phi$$
$$-\frac{1}{2}\left\{\left(\frac{a^2-b^2}{a^2+b^2}\right)^2 - \left(\frac{a-b}{a+b}\right)^2\right\}\cos 4\phi$$
$$+\frac{1}{3}\left\{\left(\frac{a^2-b^2}{a^2+b^2}\right)^3 - \left(\frac{a-b}{a+b}\right)^3\right\}\cos 6\phi$$
$$- \quad \&c. \quad \ldots\ldots\ldots\ldots\ldots\ldots \quad (D),$$

or, for the common system of logarithms,

$$\text{putting } \frac{a-b}{a+b} = n,$$

$$\log \rho = \log\left(a . \frac{1+n^2}{1+n}\right) + M\left\{\left[\left(\frac{2n}{1+n^2}\right) - n\right]\cos 2\phi\right.$$
$$-\frac{1}{2}\left[\left(\frac{2n}{1+n^2}\right)^2 - n^2\right]\cos 4\phi$$
$$+\frac{1}{3}\left[\left(\frac{2n}{1+n^2}\right)^3 - n^3\right]\cos 6\phi$$
$$\left. - \quad \&c.\right\} \quad \ldots\ldots\ldots\ldots \quad (E),$$

where M denotes the modulus of the common system of logarithms, and

$$\log M = 9{\cdot}6377843.$$

Calculating again the numerical values of the coefficients, we obtain, if a be taken $= 1$,

$$\log \rho = 9{\cdot}9992747 + 0{\cdot}0007271 \cos 2\phi - 0{\cdot}0000018 \cos 4\phi \ldots (F),$$

from which, for example, for the latitude of Berlin

$$\log \rho = 9{\cdot}9990880.$$

Consequently when the latitude of a place is known, the corrected latitude and the distance of the place from the centre of the earth can be calculated by means of the series (C) and (F), and by means of these quantities in connexion with the sidereal time the position of the place with reference to the centre of the earth can at any instant be determined. Imagine then a rectangular system of co-ordinates to be drawn through the centre of the earth, of which the axis of z is perpendicular to the plane of the equator, whilst the axes of x and y lie in the plane of the equator, and so that the positive axis of x is directed towards the first point of Aries and the positive axis of y to a point having 90° of right ascension, then can we also express the position of the place upon the surface with reference to the centre by means of the three rectangular co-ordinates,

$$\left.\begin{aligned} x &= \rho \cos \phi' \cos \theta \\ y &= \rho \cos \phi' \sin \theta \\ z &= \rho \sin \phi' \end{aligned}\right\} \quad \ldots\ldots\ldots\ldots\ldots\ldots\ldots \quad (G).$$

3. The plane, in which the lines from the centre of the earth and from the place of observation lie, passes, if the earth be considered as a sphere, necessarily through the zenith of the place of observation, and cuts consequently the visible sphere of the heavens in a vertical circle. From which it follows that the parallax alters only the altitude of the body, the azimuth remaining unchanged. Let now A (fig. 2) be the place of observation, Z its zenith, S the star, and O the centre of the earth; then ZOS is the true zenith-distance z as seen from the centre of the earth; and ZAS the apparent zenith-distance z' as observed from the place A on the surface. Denoting then the parallax, that is the angle at $S = z' - z$, by p',

$$\sin p' = \frac{\rho}{\Delta} \sin z',$$

where Δ represents the distance of the body from the earth, and, since p', except in the case of the moon, is a very small angle, the arc may be put in place of the sine, and

$$\therefore\ p' = \frac{\rho}{\Delta} \sin z' \times 206265.$$

The parallax is consequently proportional to the sine of the apparent zenith-distance. It is nothing at the zenith, attains its maximum value in the horizon, and causes the altitude of every body to appear too small. The maximum value, for $z' = 90^0$,

or
$$p = \frac{\rho}{\Delta} 206265,$$

is called the *Horizontal Parallax*, and the value

$$P = \frac{a}{\Delta} 206265,$$

where a is the radius of the earth at the equator, is called the *Horizontal-Equatoreal-Parallax.*

Hitherto the earth has been considered as spherical; but since the earth is a spheroid, the plane in which lie the lines drawn from the centre of the earth and from the place of observation to the body does not pass through the zenith of the place of observation but through the point in which the line from the centre of the earth to the place of observation meets the visible sphere of the heavens. On this account will the azimuth of the body also be altered by the parallax, and at the same time the rigorous expression for the parallax in altitude will be different from that just given*.

Let us consider then a system of three rectangular co-ordinate axes, of which the axis of z is drawn towards the zenith of the place of observation, whilst the axes of x and y lie in the plane of the horizon, and so that the positive axis of x is directed towards the south, the positive axis of y towards the west; then the co-ordinates of a body referred to these axes are

$$\Delta' \sin z' \cos A', \quad \Delta' \sin z' \sin A', \quad \text{and} \quad \Delta' \cos z',$$

where Δ' represents the distance of the body from the place of observation, z' and A' the apparent zenith-distance and azimuth as seen from the place of observation.

* The Author is indebted for the following symmetrical investigation to the kindness of Prof. Encke.

Moreover the co-ordinates of the body referred to a system of axes parallel to the former but which passes through the centre of the earth, are

$$\Delta \sin z \cos A,\ \Delta \sin z \sin A,\ \text{and}\ \Delta \cos z,$$

if we denote by Δ the distance of the body from the centre of the earth, and by z and A the zenith-distance and azimuth respectively as seen from the centre of the earth. Since now the co-ordinates of the centre of the earth referred to the first system of axes are respectively

$$-\rho \sin(\phi - \phi'),\ 0,\ \text{and}\ -\rho \cos(\phi - \phi'),$$

we obtain the three equations

$$\begin{aligned} \Delta' \sin z' \cos A' &= \Delta \sin z \cos A - \rho \sin(\phi - \phi'), \\ \Delta' \sin z' \sin A' &= \Delta \sin z \sin A, \\ \Delta' \cos z' &= \Delta \cos z - \rho \cos(\phi - \phi'); \end{aligned}$$

or,

$$\left.\begin{aligned} \Delta' \sin z' \sin(A' - A) &= \rho \sin(\phi - \phi') \sin A \\ \Delta' \sin z' \cos(A' - A) &= \Delta \sin z - \rho \sin(\phi - \phi') \cos A \\ \Delta' \cos z' &= \Delta \cos z - \rho \cos(\phi - \phi') \end{aligned}\right\} \quad (a).$$

Multiplying the first equation by $\sin\frac{1}{2}(A' - A)$, the second by $\cos\frac{1}{2}(A' - A)$, and adding the products, we obtain

$$\Delta' \sin z' = \Delta \sin z - \rho \sin(\phi - \phi') \frac{\cos\frac{1}{2}(A' + A)}{\cos\frac{1}{2}(A' - A)},$$

$$\Delta' \cos z' = \Delta \cos z - \rho \cos(\phi - \phi');$$

putting then

$$\tan\gamma = \frac{\cos\frac{1}{2}(A' + A)}{\cos\frac{1}{2}(A' - A)} \tan(\phi - \phi') \ \ldots\ldots\ldots\ldots\ldots \ (b),$$

we have

$$\Delta' \sin z' = \Delta \sin z - \rho \cos(\phi - \phi') \tan\gamma,$$

$$\Delta' \cos z' = \Delta \cos z - \rho \cos(\phi - \phi');$$

or,

$$\left.\begin{aligned}\Delta' \sin(z'-z) &= \rho\cos(\phi-\phi')\frac{\sin(z-\gamma)}{\cos\gamma}\\ \Delta'\cos(z'-z) &= \Delta-\rho\cos(\phi-\phi')\frac{\cos(z-\gamma)}{\cos\gamma}\end{aligned}\right\}\ \ldots (c).$$

And again, if we multiply the first equation by $\sin\frac{1}{2}(z'-z)$, the second by $\cos\frac{1}{2}(z'-z)$, and add the products,

$$\Delta' = \Delta - \rho\,\frac{\cos(\phi-\phi')\cos\left\{\frac{1}{2}(z'+z)-\gamma\right\}}{\cos\gamma}.$$

Dividing equations (*a*), (*b*) and (*c*) by Δ, and considering the semi-diameter of the earth at the equator as the unit, and putting

$$\frac{1}{\Delta} = \sin P,$$

where P represents the horizontal equatoreal parallax, we obtain, by means of the formulæ (14) and (15) in No. 11 of the Introduction,

$$A'-A = \frac{\rho\sin P\sin(\phi-\phi')}{\sin z}\sin A$$
$$+\frac{1}{2}\left\{\frac{\rho\sin P\sin(\phi-\phi')}{\sin z}\right\}^2\sin 2A+\ldots,$$

$$\gamma = \cos A\times(\phi-\phi') - \sin A\tan\frac{1}{2}(A'-A)\times(\phi-\phi')$$
$$+\frac{1}{3}\,\frac{\sin A\sin A'\cos\frac{1}{2}(A'+A)}{\cos^3\frac{1}{2}(A'-A)}\times(\phi-\phi')^3+\ldots\text{*},$$

* We get, namely,

$$\gamma = \frac{\cos\frac{1}{2}(A'+A)}{\cos\frac{1}{2}(A'-A)}\tan(\phi-\phi') - \frac{1}{3}\,\frac{\cos^3\frac{1}{2}(A'+A)}{\cos^3\frac{1}{2}(A'-A)}\tan^3(\phi-\phi')+\ldots$$

hence putting for $\tan(\phi-\phi')$ the series $(\phi-\phi')+\frac{1}{3}(\phi-\phi')^3+\ldots$ we obtain easily the expression given in the text.

$$z' - z = \frac{\rho \sin P \cos (\phi - \phi')}{\cos \gamma} \sin (z - \gamma)$$

$$+ \frac{1}{2} \left\{ \frac{\rho \sin P \cos (\phi - \phi')}{\cos \gamma} \right\}^2 \sin 2 (z - \gamma) + \ldots,$$

$$\log_e \Delta' = \log_e \Delta - \frac{\rho \sin P \cos (\phi - \phi')}{\cos \gamma} \cos (z - \gamma)$$

$$+ \frac{1}{2} \left\{ \frac{\rho \sin P \cos (\phi - \phi')}{\cos \gamma} \right\}^2 \cos 2 (z - \gamma) - \ldots.$$

Hence, as an approximation as far as quantities of the order $(\phi - \phi') \sin P$, which never amount to any significant value in γ,

$$\gamma = (\phi - \phi') \cos A,$$

and we obtain for the azimuthal parallax

$$A' - A = \frac{\rho \sin P \sin (\phi - \phi')}{\sin z} \sin A;$$

or, if z should be very small, accurately

$$\tan (A' - A) = \frac{\dfrac{\rho \sin P \sin (\phi - \phi')}{\sin z} \sin A}{1 - \dfrac{\rho \sin P \sin (\phi - \phi')}{\sin z} \cos A}.$$

Further, since

$$\frac{\cos (\phi - \phi')}{\cos \gamma} = \frac{\cos \frac{1}{2} (A' + A) \sin (\phi - \phi')}{\cos \frac{1}{2} (A' - A) \sin \gamma},$$

and is consequently very nearly equal to unity, we obtain for the parallax in zenith distance

$$z' - z = \rho \sin P \sin \{z - (\phi - \phi') \cos A\};$$

or, more strictly,

$$\frac{\Delta'}{\Delta} \sin (z' - z) = \rho \sin P \sin \{z - (\phi - \phi') \cos A\},$$

$$\frac{\Delta'}{\Delta} \cos (z' - z) = 1 - \rho \sin P \cos \{z - (\phi - \phi') \cos A\}.$$

Consequently, for the meridian, the parallax in azimuth is zero, and the parallax in zenith-distance is

$$z' - z = \rho \sin P \sin \{z - (\phi - \phi')\}.$$

4. Similarly we obtain the parallax for right ascension and declination. Let the co-ordinates of a body referred to the plane of the equator and the centre of the earth be

$$\Delta \cos \delta \cos \alpha, \quad \Delta \cos \delta \sin \alpha, \quad \text{and} \quad \Delta \sin \delta.$$

Let also the apparent co-ordinates referred to the same planes, as they appear from the place of observation on the surface of the earth, be

$$\Delta' \cos \delta' \cos \alpha', \quad \Delta' \cos \delta' \sin \alpha', \quad \text{and} \quad \Delta' \sin \delta'.$$

We have thus, since the co-ordinates of the station on the surface referred to the centre of the earth and to the equator as the fundamental plane, are

$$\rho \cos \phi' \cos \Theta, \quad \rho \cos \phi' \sin \Theta, \quad \text{and} \quad \rho \sin \phi',$$

the three following equations for the determination of Δ', α', and δ':

$$\left.\begin{aligned} \Delta' \cos \delta' \cos \alpha' &= \Delta \cos \delta \cos \alpha - \rho \cos \phi' \cos \Theta \\ \Delta' \cos \delta' \sin \alpha' &= \Delta \cos \delta \sin \alpha - \rho \cos \phi' \sin \Theta \\ \Delta' \sin \delta' &= \Delta \sin \delta \qquad\quad - \rho \sin \phi' \end{aligned}\right\} \ldots\ldots (a).$$

Multiplying the first equation by $\sin \alpha$, the second by $\cos \alpha$, and subtracting,

$$\Delta' \cos \delta' \sin (\alpha' - \alpha) = - \rho \cos \phi' \sin (\Theta - \alpha).$$

Again, multiplying the first equation by $\cos \alpha$, the second by $\sin \alpha$, and adding, we find

$$\Delta' \cos \delta' \cos (\alpha' - \alpha) = \Delta \cos \delta - \rho \cos \phi' \cos (\Theta - \alpha).$$

Consequently we have

$$\tan (\alpha' - \alpha) = \frac{\rho \cos \phi' \sin (\alpha - \Theta)}{\Delta \cos \delta - \rho \cos \phi' \cos (\alpha - \Theta)}$$

$$= \frac{\dfrac{\rho \cos \phi'}{\Delta \cos \delta} \sin (\alpha - \Theta)}{1 - \dfrac{\rho \cos \phi'}{\Delta \cos \delta} \cos (\alpha - \Theta)};$$

or, using the expansion which has been already frequently employed,

$$\alpha' - \alpha = \frac{\rho \cos \phi'}{\Delta \cos \delta} \sin (\alpha - \Theta) + \frac{1}{2}\left(\frac{\rho \cos \phi'}{\Delta \cos \delta}\right)^2 \sin 2 (\alpha - \Theta)$$

$$+ \frac{1}{3}\left(\frac{\rho \cos \phi'}{\Delta \cos \delta}\right)^3 \sin 3 (\alpha - \Theta)$$

$$+ \&c. \quad \ldots\ldots\ldots\ldots\ldots\ldots\ldots\ldots (A).$$

For all cases, excepting the moon, the first term of this series is sufficient, and if for ρ we take as unit the radius of the earth's equator, and then introduce into the numerator the factor $\sin \pi$ (where π is the horizontal equatoreal parallax of the sun) whereby the same unit, namely the semi-major axis of the earth's orbit, is introduced into the numerator and denominator, we get the simple equation

$$\alpha' - \alpha = \frac{\rho \sin \pi \cos \phi'}{\Delta} \cdot \frac{\sin (\alpha - \Theta)}{\cos \delta} \quad \ldots\ldots\ldots\ldots (B),$$

where $\alpha - \Theta$ is the east hour-angle of the body. The parallax consequently increases the right ascensions of stars on the east side of the meridian, and diminishes them on the west side. If the body be on the meridian, its parallax in right ascension is nothing.

In order to find a similar formula for $\delta' - \delta$, put, in the formula for $\Delta' \cos \delta' \cos (\alpha' - \alpha)$, $1 - 2 \sin^2 \frac{1}{2} (\alpha' - \alpha)$, in the place of $\cos (\alpha' - \alpha)$, by which means we obtain

$$\Delta' \cos \delta' = \Delta \cos \delta - \rho \cos \phi' \cos (\Theta - \alpha) + 2\Delta' \cos \delta' \sin^2 \frac{1}{2} (\alpha' - \alpha).$$

Multiplying and dividing the last term by

$$\cos \frac{1}{2} (\alpha' - \alpha),$$

we find, by the use of the formula

$$\Delta' \cos \delta' \sin (\alpha' - \alpha) = - \rho \cos \phi' \sin (\Theta - \alpha),$$

$$\Delta' \cos \delta' = \Delta \cos \delta - \rho \cos \phi' \frac{\cos \left\{\Theta - \frac{1}{2} (\alpha' + \alpha)\right\}}{\cos \frac{1}{2} (\alpha' - \alpha)} \quad \ldots\ldots\ldots\ldots (b).$$

Introducing now the auxiliary quantities

$$\beta \sin \gamma = \sin \phi',$$

$$\beta \cos \gamma = \frac{\cos \phi' \cos \left\{\Theta - \frac{1}{2}(\alpha' + \alpha)\right\}}{\cos \frac{1}{2}(\alpha' - \alpha)} \ldots\ldots\ldots\ldots (c),$$

we obtain from (*b*)

$$\Delta' \cos \delta' = \Delta \cos \delta - \rho\beta \cos \gamma,$$

and from the third of equations (*a*)

$$\Delta' \sin \delta' = \Delta \sin \delta - \rho\beta \sin \gamma.$$

From these two equations we easily obtain:

$$\Delta' \sin (\delta' - \delta) = -\rho\beta \sin (\gamma - \delta),$$
$$\Delta' \cos (\delta' - \delta) = \Delta - \rho\beta \cos (\gamma - \delta),$$

or

$$\tan (\delta' - \delta) = -\frac{\frac{\rho\beta}{\Delta} \sin (\gamma - \delta)}{1 - \frac{\rho\beta}{\Delta} \cos (\gamma - \delta)},$$

or also, according to the formula (14) in No. 11 of the Introduction,

$$\delta' - \delta = -\frac{\rho\beta}{\Delta} \sin (\gamma - \delta) - \frac{1}{2}\frac{\rho^2\beta^2}{\Delta^2} \sin 2(\gamma - \delta) - \&c. \ldots (C).$$

Putting here for β its value $\frac{\sin \phi'}{\sin \gamma}$, and $\rho \sin \pi$ instead of ρ, in order to have the same unit in the numerator and denominator, we obtain, keeping only the first term of the series,

$$\delta' - \delta = -\frac{\rho \sin \pi \sin \phi'}{\Delta} \cdot \frac{\sin (\gamma - \delta)}{\sin \gamma}.$$

Putting, moreover, in the second of formulæ (*c*) (which is always allowable) unity instead of $\cos \frac{1}{2}(\alpha' - \alpha)$, and α instead of $\cos \frac{1}{2}(\alpha' + \alpha)$, the complete approximate formulæ for the computation of the parallax in right ascension and declination are the following:

$$\alpha' - \alpha = -\frac{\pi\rho \cos \phi'}{\Delta} \cdot \frac{\sin (\Theta - \alpha)}{\cos \delta},$$

$$\tan\gamma = \frac{\tan\phi'}{\cos(\Theta - \alpha)},$$

$$\delta' - \delta = -\frac{\pi\rho\sin\phi'}{\Delta} \cdot \frac{\sin(\gamma - \delta)}{\sin\gamma} \text{*}.$$

If the body have a visible disk, its apparent semi-diameter will depend on the distance, and a correction will still be necessary on this account. Now we have

$$\Delta' \sin(\delta' - \gamma) = \Delta \sin(\delta - \gamma);$$

$$\therefore\ \Delta' = \Delta \cdot \frac{\sin(\delta - \gamma)}{\sin(\delta' - \gamma)},$$

and, since the semi-diameters, as long as they are small angles, vary inversely as the distances, we have

$$R' = R \cdot \frac{\sin(\delta' - \gamma)}{\sin(\delta - \gamma)}.$$

Example.

1844, Sep. 3, at $20^h. 41^m. 38^s$ sidereal, a comet, discovered by de Vico, was observed

in right ascension $2^0.35'. 55'',5$,

and in declination $-18^0.43'. 21'',6$.

The logarithm of the distance from the earth was at this time 9·28001, and in addition for Rome

$$\phi' = 41^0. 42',5,$$

and $\log\rho = 9{\cdot}99936.$

With these data the calculation for the parallax is the following:

Θ in arc	=	$310^0. 24',5$
α	=	$2 . 35,9$
$\Theta - \alpha$	=	$-52 . 11,4$

* For the meridian we obtain from this formula,

$$\delta' - \delta = -\frac{\pi\rho}{\Delta}\sin(\phi' - \delta) = \rho \cdot \frac{\pi}{\Delta} \cdot \sin[z - (\phi - \phi')],$$

consequently the parallax in declination is equal to the parallax in altitude.

$$\begin{aligned}\log\tan\phi' &= 9{\cdot}94999\\ \log\cos(\Theta-\alpha) &= 9{\cdot}78749\\ \log\sin(\Theta-\alpha) &= -9{\cdot}89765\\ -\log\frac{\pi\rho\cos\phi'}{\Delta} &= -1{\cdot}52576\\ \log\sec\delta &= 0{\cdot}02362\\ \log(\alpha'-\alpha) &= 1{\cdot}44703\\ \alpha'-\alpha &= +27'',99\end{aligned}$$

$$\begin{aligned}\gamma &= 55^0.28',6\\ \delta &= -18\,.43\,,4\\ \gamma-\delta &= 74\,.12\,,0\\ \log\sin(\gamma-\delta) &= +9{\cdot}98327\\ -\log\frac{\pi\rho\sin\phi'}{\Delta} &= -1{\cdot}47576\\ \log\operatorname{cosec}\gamma &= 0{\cdot}08413\\ \log(\delta'-\delta) &= -1{\cdot}54316\\ \delta'-\delta &= -34'',93\end{aligned}$$

On account of the parallax the observed right ascension of the comet is consequently greater by 28″,0, and the declination is less by 34″,9, than these quantities would have been if observed from the centre of the earth. The place of the comet freed from the parallax is consequently

$$\begin{aligned}\alpha &= 2^0.\,35'\,.27'',5,\\ \delta &= -18^0.\,42'.\,46'',7.\end{aligned}$$

To obtain the parallax of an object for co-ordinates which are referred to the ecliptic as the fundamental plane, it is necessary to know the co-ordinates of the place of observation with reference to the centre of the earth, for the same fundamental plane. But by changing Θ and ϕ' into longitude and latitude according to No. 8 of the first section, and for this purpose obtaining the values of l and b, these co-ordinates are:

$$\rho\cos b\cos l,$$
$$\rho\cos b\sin l,$$
$$\rho\sin b,$$

and we have, if λ', β', Δ' be the apparent and λ, β, Δ the true quantities, the three equations

$$\begin{aligned}\Delta'\cos\beta'\cos\lambda' &= \Delta\cos\beta\cos\lambda-\rho\cos b\cos l,\\ \Delta'\cos\beta'\sin\lambda' &= \Delta\cos\beta\sin\lambda-\rho\cos b\sin l,\\ \Delta'\sin\beta' &= \Delta\sin\beta \qquad\quad -\rho\sin b,\end{aligned}$$

from whence we obtain formulæ similar to those obtained before, namely:

$$\lambda'-\lambda=-\frac{\pi\rho\cos b}{\Delta}\cdot\frac{\sin(l-\lambda)}{\cos\beta},$$

$$\tan\gamma = \frac{\tan b}{\cos(l-\lambda)},$$

$$\beta' - \beta = -\frac{\pi\rho\sin b}{\Delta}\cdot\frac{\sin(\gamma-\beta)}{\sin\gamma}.$$

Θ and ϕ' are the right ascension and declination of that point in which the radius of the earth produced meets the sphere of the heavens; l and b are consequently the longitude and latitude of this same point. If we consider the earth as spherical, this point coincides with the zenith, and its longitude is also called the *Nonagesimal*, since the point of the ecliptic corresponding to this longitude is distant 90° from that point of it which is in the horizon.

5. Since the horizontal parallax of the moon or the angle $\frac{\rho\sin P}{\Delta}$, where Δ denotes the distance of the moon from the earth, lies always between 54′ and 61′, it follows that, for the computation of the moon's parallax the first terms of the series for $\alpha'-\alpha$ and $\delta'-\delta$ will not be sufficient, but that it will be necessary either to take account of the terms of higher orders, or to use the rigorous formula.

Suppose the parallax of the moon in right ascension and declination to be required for Greenwich on 1848, April 10, at 10^h.

For this time we have

$$\alpha = 7^h.\,43^m.\,20^s{,}25 = 115^\circ.\,50'.\,3''{,}75,$$
$$\delta = +16^\circ.\,27'.\,22''{,}9,$$
$$\Theta = 11^h.\,17^m.\,0^s{,}02 = 169^\circ.\,15'.\,0''{,}30,$$

the horizontal parallax

$$P = 56'.\,57''{,}5,$$
$$R = 15'.\,31''{,}3,$$

in addition, we have for Greenwich

$$\phi' = 51^\circ.\,17'.\,25''{,}4,$$
$$\log\rho = 9{\cdot}9991134.$$

By introducing the horizontal parallax P of the moon into the two series for $\alpha'-\alpha$, and $\delta'-\delta$, found in No. 4, we shall have

$$\alpha' - \alpha = -206265\left\{\frac{\rho\cos\phi'\sin P}{\cos\delta}\sin(\Theta-\alpha)\right.$$

$$+\frac{1}{2}\left(\frac{\rho\cos\phi'\sin P}{\cos\delta}\right)^2\sin 2(\Theta-\alpha)$$

$$+\frac{1}{3}\left(\frac{\rho\cos\phi'\sin P}{\cos\delta}\right)^3\sin 3(\Theta-\alpha)+\ldots\Big\}.$$

And

$$\delta'-\delta=-206265\left\{\frac{\rho\sin\phi'\sin P}{\operatorname{cosec}\gamma}\sin(\gamma-\delta)\right.$$

$$+\frac{1}{2}\left(\frac{\rho\sin\phi'\sin P}{\operatorname{cosec}\gamma}\right)^2\sin 2(\gamma-\delta)$$

$$\left.+\frac{1}{3}\left(\frac{\rho\sin\phi'\sin P}{\operatorname{cosec}\gamma}\right)^3\sin 3(\gamma-\delta)+\ldots\right\},$$

where we must now, for the computation of the auxiliary angle γ, employ the rigorous formula

$$\tan\gamma=\tan\phi'\cdot\frac{\cos\frac{1}{2}(\alpha'-\alpha)}{\cos\left\{\Theta-\frac{1}{2}(\alpha'+\alpha)\right\}}.$$

Performing then the computations for these formulæ, we thus obtain for $\alpha'-\alpha$

	from the first term	$-29'.45'',71$
	.... second...	$-\quad 11,47$
	.... third ...	$-\quad 0,03$
thus	$\alpha'-\alpha=$	$-29.57,21$

and, for $\delta'-\delta$;

	from the first term	$-36'.34'',21$
	.... second ...	$-\quad 20,91$
	.... third ...	$-\quad 0,12$
thus	$\delta'-\delta=$	$-36.55,24$

The apparent right ascension and declination of the moon is therefore

$$\alpha'=115^\circ.20'.6'',54,\quad \delta'=+15^\circ.50'.27'',66.$$

Lastly, we obtain for the apparent semi-diameter

$$R'=15'.40'',20.$$

If it be preferred to compute the parallax by the rigorous formula, it must be rendered more convenient for logarithmic computation. The rigorous formula for $\tan(\alpha' - \alpha)$ was

$$\tan(\alpha' - \alpha) = \frac{\rho \cos\phi' \sin P \sin(\alpha - \Theta) \sec\delta}{1 - \rho\cos\phi' \sin P \cos(\alpha - \Theta)\sec\delta} \;\ldots\ldots\; (a).$$

In addition, it follows from the two equations

$$\Delta' \sin\delta' = \sin\delta - \rho\sin\phi' \sin P,$$

and

$$\Delta' \cos\delta' \cos(\alpha' - \alpha) = \cos\delta - \rho\cos\phi' \sin P \cos(\alpha - \Theta),$$

that $\tan\delta' = \dfrac{[\sin\delta - \rho\sin\phi' \sin P]\cos(\alpha' - \alpha)\sec\delta}{1 - \rho\cos\phi' \sin P \sec\delta\cos(\alpha - \Theta)} \;\ldots\ldots\; (b).$

Since also

$$\frac{\Delta}{\Delta'} = \frac{\cos\delta' \cos(\alpha' - \alpha)}{\cos\delta - \rho\cos\phi' \sin P\cos(\alpha - \Theta)},$$

we thus obtain in addition

$$\sin R' = \frac{\cos\delta' \cos(\alpha' - \alpha)\sec\delta\sin R}{1 - \rho\cos\phi' \sin P\sec\delta\cos(\alpha - \Theta)} \;\ldots\ldots\; (c).$$

Introduce now into (a), (b), and (c) the auxiliary quantities

$$\cos A = \frac{\rho\sin P\cos\phi' \cos(\alpha - \Theta)}{\cos\delta},$$

and

$$\sin C = \rho\sin P\sin\phi',$$

and we thus obtain the following convenient formulæ for logarithmic computation,

$$\tan(\alpha' - \alpha) = \frac{\frac{1}{2}\rho\cos\phi' \sin P\sin(\alpha - \Theta)}{\cos\delta\sin^2\frac{1}{2}A},$$

$$\tan\delta' = \frac{\sin\frac{1}{2}(\delta - C)\cos\frac{1}{2}(\delta + C)\cos(\alpha' - \alpha)}{\cos\delta\sin^2\frac{1}{2}A},$$

and

$$R' = \frac{\frac{1}{2}\cos\delta' \cos(\alpha' - \alpha)}{\cos\delta\sin^2\frac{1}{2}A} \cdot R.$$

If $\alpha' - \alpha$, δ', and R' be required for the preceding example according to these formulæ, we obtain with precisely the same accuracy as before,

$$\alpha' - \alpha = - \quad 29'.57'',21,$$
$$\delta = + 15^0.50'.27'',68,$$
$$R' = \quad 15'.40'',21.$$

For the rigorous computation of the parallaxes in longitude and latitude we obtain precisely similar formulæ, with this sole difference that λ', λ, β', β, l, and b appear in place of α', α, δ', δ, Θ, and ϕ'.

II. Refraction.

6. The rays of light do not reach us through empty space, but through the atmosphere of the earth. In empty space the rays proceed in straight lines; but when they enter into another medium, which refracts the light, they are turned aside from their original direction. If now this medium consists of innumerable layers whose refractive powers continuously vary as is the case with our atmosphere, the path of the ray of light through it will be a continuous curve. An observer on the earth now sees the object in the direction of the last tangent to the curve, which the ray describes, and he must from this direction, which defines the apparent place of the object, determine that direction of the ray which it would have had in empty space, that is, the true position of the object. The difference between these two directions is called *Refraction*, and, since the curve formed by the path of the ray in the atmosphere turns its concave side towards the observer, it is evident that, on account of refraction, all objects are seen at too great an altitude.

In what follows the shape of the earth will be supposed to be spherical, since the effect of the spheroidal figure of the earth on the refraction is altogether insignificant. The atmosphere will be supposed to consist of concentric layers, within which the density, and consequently the refraction which depends on it, is constant. Now for the purpose of estimating the change of direction of the rays in each layer on account of refraction, it is necessary to know the laws of the refraction of light. These are four, namely the following:

(1) When a ray of light falls on any surface of a body, which separates two media of different refractive powers, if we draw a tangent plane at the point where the ray of light is incident, a normal to this plane, and finally a plane through it and the path of the ray, the ray will not deviate from this plane after its entrance into the body.

(2) If we suppose the normal to be produced outwards, then for any media whatever and for all angles of incidence, the sine of the angle of incidence (that is, of the angle between the incident ray and the normal) will bear a constant ratio to the sine of the angle of refraction (that is, of the angle between the refracted ray and the normal). This ratio is called the *Refractive Index* for these two media.

(3) When the refractive index between two media A and B is given, and also that between two other media B and C, the refractive index between the media A and C is the compound ratio of the indices between A and B and between B and C.

(4) If μ be the refractive index for the passage from one medium A into another B, then is $\frac{1}{\mu}$ the refractive index for the passage from the medium B into the medium A.

Let now O (fig. 3) be a place on the surface of the earth, C the earth's centre, S the true place of a star, CJ the normal to the point J, at which the ray of light SJ meets the first layer of the atmosphere. Then if the refractive index for this first layer be known, we can by the laws of refraction find the direction of the refracted ray, and we obtain for the second layer a new angle of incidence. Suppose now that there are n layers, and suppose CN to be the line from the centre of the earth to the point, at which the ray of light meets the nth layer; let also i_n be the angle of incidence, f_n the angle of refraction, μ_n the refractive index from vacuum to the nth, and μ_{n+1} that to the $n+1^{\text{th}}$ layer; then we have*

* These refractive indices are fractions, whose numerators are greater than unity. For layers at the surface of the earth for example $\mu = 1{\cdot}000294$, or nearly $\frac{3400}{3399}$.

$$\frac{\sin i_n}{\sin f_n} = \frac{\mu_{n+1}}{\mu_n}.$$

If then N' be the point at which the ray of light meets the $n+1^{\text{th}}$ layer, we have in the triangle NCN', denoting by r_n and r_{n+1} the distances of the points N and N' from the centre of the earth,

$$\frac{\sin f_n}{\sin i_{n+1}} = \frac{r_{n+1}}{r_n},$$

and, combining this equation with the former,

$$r_n \sin i_n \, \mu_n = r_{n+1} \sin i_{n+1} \, \mu_{n+1}.$$

Consequently, since the product of the distance from the centre into the refractive index and the sine of the angle of incidence is the same for all strata of the atmosphere, we thus obtain, if we denote this constant by α_1, as the general law of refraction

$$r\mu \sin i = \alpha_1 \quad \ldots\ldots\ldots\ldots\ldots\ldots\ldots \quad (a),$$

where r, μ and i must belong to the same point of the atmosphere. For the surface of the earth, i, (that is the angle which the last tangent of the ray makes with the normal), will be equal to the apparent zenith-distance of the star. Thus, if we call a the radius of the earth, and μ_0 the refractive index for a stratum of the atmosphere at its surface, we obtain for the determination of the constant α_1, the equation

$$a\mu_0 \sin z = \alpha_1 \quad \ldots\ldots\ldots\ldots\ldots\ldots\ldots \quad (b).$$

Assume now that the density of the atmosphere varies continuously, and that consequently the altitude of the stratum within which the density may be regarded as constant is indefinitely small, then the path of the ray through the atmosphere will be a curve, whose equation can be determined. Introducing polar co-ordinates, and calling v the angle which any value of r makes with the radius CO, we easily obtain

$$r\frac{dv}{dr} = \tan i \quad \ldots\ldots\ldots\ldots\ldots\ldots\ldots \quad (c).$$

The direction of the last tangent is, as has just been seen, the apparent zenith-distance z, whilst the true zenith-distance ζ

is the angle which the original direction SJ of the ray produced, makes with the normal. This ζ has, it is true, its vertex at another point from that at which the eye of the observer is situated, but since the atmosphere is only of small elevation, and on the contrary the luminous bodies are very distant, and especially since the refraction itself is only a small angle, the difference between the angle ζ and the true zenith-distance, which is to be observed at O, is quite insignificant. Even in the case of the moon, in which this difference is most conspicuous, it amounts only to a very small part of a second. We may therefore assume that the angle ζ is the true zenith-distance.

At the point N, to which the variable quantities i, r, and μ apply, draw now a tangent to the ray, which makes with the normal CO the angle ζ'; then

$$\zeta' = i + v \qquad \text{(d)}.$$

Differentiating then the general equation (a) logarithmically, we obtain

$$\frac{dr}{r} + \operatorname{cotan} i \,.\, di + \frac{d\mu}{\mu} = 0,$$

and from this equation, in connection with the equations (c) and (d),

$$d\zeta' = -\tan i \frac{d\mu}{\mu};$$

or, if we eliminate $\tan i$ by the equation

$$\tan i = \frac{\sin i}{\sqrt{(1-\sin^2 i)}} = \frac{\alpha_1}{\sqrt{(r^2\mu^2 - \alpha_1^2)}},$$

and put for α_1 its value $a\mu_0 \sin z$,

$$d\zeta' = -\frac{\frac{a}{r}\mu_0 \sin z \, d\mu}{\mu\sqrt{\left(\mu^2 - \frac{a^2}{r^2}\mu_0^2 \sin^2 z\right)}} \qquad \text{(e)}.$$

The integral of this equation, taken between the limits

$\zeta' = \zeta$ and $\zeta' = z$, gives then the amount of refraction. If we put

$$\frac{a}{r} = 1 - s,$$

we can also write the equation thus,

$$d\zeta' = -\frac{(1-s)\sin z\, d\mu}{\mu\sqrt{\left\{\cos^2 z - \left(1 - \frac{\mu^2}{\mu_0^2}\right) + (2s - s^2)\sin^2 z\right\}}} \quad \dots (f).$$

To integrate this equation, we must now know the value of s as a function of μ. This latter quantity is dependent on the density, and we are taught by the physical sciences that the quantity $\mu^2 - 1$, which is also called the *refractive power*, is proportional to the density. Introducing then the density ρ as a new variable, given by the equation

$$\mu^2 - 1 = c\rho,$$

where c is a constant, we obtain

$$* \; d\zeta' = \frac{\frac{1}{2}(1-s)\sin z \,.\, c \,.\, \frac{d\rho}{1+c\rho_0}}{\frac{1+c\rho}{1+c\rho_0}\,.\sqrt{\left\{\cos^2 z - \left(1 - \frac{1+c\rho}{1+c\rho_0}\right) + (2s - s^2)\sin^2 z\right\}}},$$

or, if we put

$$\frac{c\rho_0}{1+c\rho_0} = \frac{\mu_0^2 - 1}{\mu_0^2} = 2\alpha,$$

$$d\zeta' = \frac{\alpha(1-s)\sin z\,\frac{d\rho}{\rho_0}}{\left\{1 - 2\alpha\left(1 - \frac{\rho}{\rho_0}\right)\right\}.\sqrt{\left\{\cos^2 z - 2\alpha\left(1 - \frac{\rho}{\rho_0}\right) + (2s - s^2)\sin^2 z\right\}}} \quad \dots\dots\dots (g).$$

The coefficient

$$1 - 2\alpha\left(1 - \frac{\rho}{\rho_0}\right)$$

* ρ_0 is the value of ρ corresponding to $\mu = \mu_0$.

Since also $\mu^2 = 1 + c\rho$, $2\log_e \mu = \log_e(1 + c\rho)$;

$$\therefore \frac{d\mu}{\mu} = \tfrac{1}{2} \cdot \frac{c\,d\rho}{1+c\rho},$$

whence, by an easy transformation, we get the form of the equation given in the text. [TRANSLATOR.]

is the square of the ratio of the index of refraction for a stratum at radius r to the index of refraction for a stratum at the surface of the earth. But since, for the limits of the atmosphere, $\mu = 1$, while, on the contrary, for the refraction out of vacuum into strata at the surface of the earth $\mu_0 = \frac{3400}{3900}$, the ratio $\frac{\mu}{\mu_0}$ always lies between these narrow limits. Consequently the quantity α is small, and instead of the variable factor

$$1 - 2\alpha\left(1 - \frac{\rho}{\rho_0}\right),$$

we can take its mean value between the two extreme limits 1 and $1 - 2\alpha$, namely, the constant value $1 - \alpha$.

In order to be able to integrate the equation (g), s must be expressed as a function of ρ, that is, we must determine the law, according to which the density of the atmosphere varies with the height above the surface of the earth. Considering first the temperature of the atmosphere as uniform, the density becomes a simple function of the pressure or of the elasticity of the air, and we have according to Mariott's law, if p denote the pressure of the air at a point whose distance from the centre of the earth is r,

$$p = p_0 \frac{\rho}{\rho_0}.$$

If then r be increased by dr the decrement of pressure is equal to the small column ρdr, and if this be multiplied into g, the force of gravity corresponding to the distance r, we obtain

$$dp = -g\rho\, dr,$$

or, since also

$$g = g_0 \frac{a^2}{r^2},$$

where g_0 represents the force of gravity at the surface of the earth;

$$dp = -g_0 \frac{a^2}{r^2} \rho\, dr,$$

consequently also

$$p_0 \frac{d\rho}{\rho_0} = g_0 a \rho d\left(\frac{a}{r}\right).$$

Integrating this equation, and remembering, in the determination of the constant, that for $\rho = \rho_0$, $r = a$, we obtain

$$\rho = \rho_0 e^{\left(\frac{a}{r} - 1\right) a \frac{\rho_0 g_0}{p_0}},$$

where e is the base of the natural system of logarithms. Taking then l for the height of a column of air of the density ρ_0* which corresponds to the force of gravity g_0 and to the pressure p_0, equilibrium holds when

$$p_0 = g_0 \rho_0 l,$$

we finally obtain, if we again put

$$\frac{a}{r} = 1 - s,$$

$$\rho = \rho_0 e^{-\frac{as}{l}}.$$

This equation gives thus for every value of s and therefore for every value of r the density of the air on the supposition that the temperature is uniform throughout the whole atmosphere. Since this supposition does not correspond to the case of nature, because the temperature of the atmosphere decreases with the elevation according to an unknown law, it becomes necessary to make some hypothesis concerning the law according to which the density of air varies.

Bessel, to whom we owe the most accurate Refraction Tables, adopted for this law the following expression

$$\rho = \rho_0 e^{-\frac{h-l}{h} \cdot \frac{as}{l}},$$

where h is a constant, which must be determined in such a manner that the refractions calculated according to this law shall correspond to the observed refractions. Putting then

$$\frac{h-l}{hl} a = \beta \qquad \ldots\ldots\ldots\ldots\ldots\ldots (h),$$

and replacing, in equation (g), ρ by its value given by the formula

$$\rho = \rho_0 e^{-\beta s} \qquad \ldots\ldots\ldots\ldots\ldots\ldots (i),$$

* This l is, for the temperature of 8° Reaumur or 50° Fahrenheit, equal to 4226·05 toises. It is equal to the mean height of the barometer at the level of the sea multiplied into the density of mercury relatively to that of air.

we obtain

$$d\zeta' = + \frac{\alpha\beta e^{-\beta s} \sin z \,(1-s)\, ds}{(1-\alpha)\sqrt{\{\cos^2 z - 2\alpha\,(1-e^{-\beta s}) + (2s - s^2)\sin^2 z\}}};$$

or, if the quantity under the radical sign be expanded in powers of s, while $-s^2 \sin^2 z$ is considered as a small increase of the remaining terms under the radical;

$$d\zeta' = \frac{\alpha\beta e^{-\beta s} \sin z\, ds}{(1-\alpha)\,\{\cos^2 z - 2\alpha\,(1-e^{-\beta s}) + 2s\sin^2 z\}^{\frac{1}{2}}}$$
$$- \frac{\alpha\beta s\, ds\, e^{-\beta s} \sin z\,\{\cos^2 z - 2\alpha\,(1-e^{-\beta s}) + s\sin^2 z\}}{(1-\alpha)\,\{\cos^2 z - 2\alpha\,(1-e^{-\beta s}) + 2s\sin^2 z\}^{\frac{3}{2}}} \;\ldots\ldots\; (k)$$
$$- \quad \&c.$$

7. Integrating the equation (k) for s between the limits $s = H$, where H is the height of the atmosphere, and $s = 0$, we obtain the value of the refraction. In order to obtain however the positive sign, the limits in what follows are to be taken in the inverse order, so that the value then found for the refraction is to be so understood, that we must add it algebraically to the apparent place in order to obtain the mean place.

Since now, from the height of the atmosphere being very small compared with the radius of the earth, s is always a small quantity, the first term in the series for $d\zeta'$ will be considerably greater than those which follow, which have only a slight influence, and it can be easily demonstrated that the second term is so small, that it may be always neglected. This term in fact attains its greatest value when $z = 90^0$, and consequently the observed object is in the horizon, viz.

$$-\frac{\alpha\beta s\, ds\, e^{-\beta s}\,\{s - 2\alpha\,(1-e^{-\beta s})\}}{(1-\alpha)\,\{2s - 2\alpha\,(1-e^{-\beta s})\}^{\frac{3}{2}}}.$$

In order to integrate this expression, we must expand it in a series of powers of s. But it must be borne in mind that the most important part of the integral is that where s is very small, and that thus the first term only is required to be taken, and we obtain then (since $1 - e^{-\beta s} = \beta s$)

$$-\frac{\alpha\beta\sqrt{s}\,.\,ds\, e^{-\beta s}\,(1 - 2\alpha\beta)}{(1-\alpha)\; 2^{\frac{3}{2}}\,(1-\alpha\beta)^{\frac{3}{2}}}.$$

We should have now to integrate this expression for particular values from $s=0$ to $s=H$, but we may without sensible error take also for these limits 0 and ∞, and we shall then find for the integral, observing that

$$\int_0^\infty \sqrt{x}\,.\,e^{-x}\,dx = \frac{1}{2}\sqrt{\pi}\,*,$$

the following value:

$$-\frac{\alpha\,(1-2\alpha\beta)\sqrt{\left(\frac{\pi}{2\beta}\right)}}{4\,(1-\alpha)\,(1-\alpha\beta)^{\frac{3}{2}}}.$$

Substituting in this the numerical value of the constant found farther on, we obtain the result 0″,55, and since this is the maximum value of the integral of the second term, which besides only applies to the horizon, this term and *à fortiori* the following terms may be neglected.

The expression is thus reduced to the first term only of the equation (*k*), viz.

$$d\zeta' = \frac{\alpha\beta\,e^{-\beta s}\sin z\,ds}{(1-\alpha)\,\{\cos^2 z - 2\alpha\,(1-e^{-\beta s}) + 2s\sin^2 z\}^{\frac{3}{2}}} \;\ldots\ldots\; (l).$$

Introducing the new variable

$$s' = -\frac{\alpha\,(1-e^{-\beta s})}{\sin^2 z} + s,$$

the denominator becomes simply

$$(1-\alpha)\,(\cos^2 z + 2s'\sin^2 z)^{\frac{3}{2}},$$

and we have only then to express $e^{-\beta s}ds$ in terms of s'. But since

$$s = s' + \alpha\,.\,\frac{1-e^{-\beta s}}{\sin^2 z} \;\ldots\ldots\ldots\ldots\ldots\ldots\ldots\; (m),$$

we can expand the quantity $e^{-\beta s}$, by substituting for s the value from this equation, in powers of α. Putting for instance $e^{-\beta s} = u$, we have by Maclaurin's theorem

$$u = U + \alpha q + \alpha^2 q + \ldots + \alpha^n q_n + \ldots \;\ldots\ldots\ldots\; (\alpha),$$

* This integral is a function represented in analysis by Γ, called Euler's integral and $\Gamma\left(\frac{3}{2}\right)$.

where U is the value of u when $\alpha = 0$, or in this case $= e^{-\beta s'}$,

and $q_n = \dfrac{\dfrac{d^n u}{d\alpha^n}}{1.2.3\ldots n}$ when $\alpha = 0$.

We thus need only expand the value $\dfrac{d^n u}{d\alpha^n}$ when $\alpha = 0$ from equation (m). But writing this equation thus*,

$$s = t + \alpha y,$$

we have

$$\left(\frac{ds}{d\alpha}\right) = y\left(\frac{ds}{dt}\right);$$

and also

$$\left(\frac{du}{d\alpha}\right) = \left(\frac{du}{ds}\right)\left(\frac{ds}{d\alpha}\right) = y\left(\frac{du}{ds}\right)\left(\frac{ds}{dt}\right) = y\left(\frac{du}{dt}\right) \ldots\ldots\ldots (\beta).$$

If now we integrate $y du$ with respect to t and afterwards again differentiate with respect to t, whereby the value of $y du$ will not be altered, we obtain

$$\left(\frac{du}{d\alpha}\right) = \left(\frac{d\int y\,du}{dt}\right),$$

and consequently

$$\left(\frac{d^2u}{d\alpha^2}\right) = \left(\frac{d^2\int y\,du}{d\alpha\,dt}\right).$$

But there follows from equation (β), if we put $\int y\,du$ instead of u,

$$\left(\frac{d\int y\,du}{d\alpha}\right) = \left(\frac{d\int y^2\,du}{dt}\right),$$

thus

$$\left(\frac{d^2u}{d\alpha^2}\right) = \left(\frac{d^2\int y^2\,du}{dt^2}\right);$$

similarly we obtain

$$\left(\frac{d^3u}{d\alpha^3}\right) = \left(\frac{d^3\int y^3\,du}{dt^3}\right),$$

and generally

$$\left(\frac{d^nu}{d\alpha^n}\right) = \left(\frac{d^n\int y^n\,du}{dt^n}\right) = \frac{d^{n-1}\left\{y^n\left(\frac{du}{dt}\right)\right\}}{dt^{n-1}} \ldots\ldots\ldots\ldots\ldots\ldots (\gamma).$$

* What follows is nothing but Laplace's investigation of his Theorem, and might have been omitted.—TRANSLATOR.

And since now for $\alpha = 0$

$$u = e^{-\beta s'},$$

$$y^n = \left\{\frac{1 - e^{-\beta s'}}{\sin^2 z}\right\}^n \ldots\ldots\ldots\ldots\ldots\ldots\ldots\ldots (\delta);$$

we at last obtain by combining equations (α), (γ), and (δ),

$$\left.\begin{aligned} e^{-\beta s} = e^{-\beta s'} &- \frac{\alpha\beta}{\sin^2 z}\left(1 - e^{-\beta s'}\right) e^{-\beta s'} \\ &- \frac{\alpha^2\beta}{1\,.\,2\,\sin^4 z}\,\frac{d.\{(1 - e^{-\beta s'})^2 e^{-\beta s'}\}}{ds'} \\ &- \&c.\ \ldots\ldots\ldots\ldots \\ &- \frac{\alpha^n\beta}{1\,.\,2\,.\,3\ldots n\,\sin^n z}\,.\,\frac{d^{n-1}\{(1 - e^{-\beta s'})^n e^{-\beta s'}\}}{ds'^{n-1}} \end{aligned}\right\} \ldots\ldots\ldots (\epsilon).$$

In the equation (l) however the numerator was to be expressed in terms of the new variable s'. But since

$$\beta ds e^{-\beta s} = -d\,.\,e^{-\beta s},$$

the equation (l) will thus become, if we expand $d\,.\,e^{-\beta s}$ by equation (ϵ),

$$d\zeta' = \frac{\alpha\beta\sin z\,ds'}{(1-\alpha)(\cos^2 z + 2s'\sin^2 z)^{\frac{1}{2}}} \left\{\begin{aligned} & e^{-\beta s'} \\ & - \frac{\alpha}{\sin^2 z}\,\frac{d.\{(e^{-\beta s'} - 1)\,e^{-\beta s'}\}}{ds'} \\ & + \frac{\alpha^2}{1\,.\,2\ldots\sin^4 z}\,\frac{d^2\{(e^{-\beta s'} - 1)^2 e^{-\beta s'}\}}{ds'^2} \\ & - \ldots\ldots\ldots\ldots\ldots \\ & \pm \frac{\alpha^n}{1\,.\,2\,.\,3\ldots n\sin^{2n} z}\,\frac{d^n\{(e^{-\beta s'} - 1)^n e^{-\beta s'}\}}{ds'^n} \\ & \mp \&c.\ldots\ldots\ldots\ldots \end{aligned}\right\} (n),$$

where in the general term the upper or lower sign must be taken according as n is even or odd.

But we have

$$\pm \frac{\alpha^n}{1\,.\,2\,.\,3\ldots n\,\sin^{2n} z}\,.\,\frac{d^n\{(e^{-\beta s'} - 1)^n e^{-\beta s'}\}}{ds'^n}$$

$$= \frac{\alpha^n \beta^n}{1.2.3 \ldots n \sin^{2n} z} \left\{ \begin{array}{l} (n+1)^n e^{-(n+1)\beta s'} \\ - n \, . \, n^n e^{-n\beta s'} \\ + \frac{n(n-1)}{1.2} (n-1)^n e^{-(n-1)\beta s'} \\ - \ldots\ldots\ldots\ldots \end{array} \right\},$$

as is evident, if we expand

$$(e^{-\beta s'} - 1)^n$$

in a series, multiply every term by $e^{-\beta s'}$, and then differentiate n times with respect to s'; accordingly we obtain, if we substitute in equation (n) for the separate terms, this series:

$$d\zeta = \frac{\alpha}{1-\alpha} \cdot \frac{\beta \sin z ds'}{(\cos^2 z + 2s' \sin^2 z)^{\frac{3}{2}}} \left\{ \begin{array}{l} e^{-\beta s'} \\ + \frac{\alpha\beta}{\sin^2 z} (2e^{-2\beta s'} - e^{-\beta s'}) \\ + \frac{\alpha^2 \beta^2}{1.2 \sin^4 z} (3^2 e^{-3\beta s'} - 2 \, . \, 2^2 e^{-\beta s'} + e^{-\beta s'}) \\ + \frac{\alpha^3 \beta^3}{1.2.3 \sin^6 z} (4^3 e^{-4\beta s'} - 3 \, . \, 3^3 e^{-3\beta s'} \\ \qquad\qquad + \frac{3.2}{1.2} 2^3 e^{-2\beta s'} - e^{-\beta s'}) \\ + \&c. \ldots\ldots\ldots\ldots\ldots\ldots\ldots\ldots\ldots \end{array} \right\} \ldots (o);$$

and this equation has now to be integrated between the proper limits. And the integral must always be taken from the surface of the earth to the boundary of the atmosphere, whose height we suppose to be H, and therefore from $r = a$ to $r = a + H$. But since at the boundary of the atmosphere the density is nothing, and consequently the ray of light, if we consider it as proceeding from the eye of the observer, after reaching that point, experiences no further refraction, we can therefore take for the limit also $r = a$ and $r = \infty$; and since

$$s = 1 - \frac{a}{r},$$

the limits, in relation to these variables, will therefore become $s = 0$ and $s = 1$, and with reference to s', will be $s' = 0$, and

$s' = 1 - \frac{\alpha(1 - e^{-\beta})}{\sin^2 z}$. But, following out equation (ι), since β is a very small number, while at the boundary of the atmosphere the density is nothing, the formula must therefore at least give so small a value for ρ, that we may neglect it without any sensible error. The quantity $e^{-n\beta s'}$ thus becomes, if we substitute the superior limit for s', collectively very small, and we can therefore take in the integral of equation (o) without sensible error, $s' = 0$ and $s' = \infty$ as the limits.

The terms of equation (o) include now, each one of them, a factor of the form

$$\frac{\beta ds' e^{-r\beta s'} \sin z}{\sqrt{(\cos^2 z + 2s' \sin^2 z)}} \quad \text{.........................} (\eta).$$

Introducing here a new variable t, obtained from the equation

$$\frac{1}{2}\frac{\cos^2 z}{\sin^2 z} + s' = \frac{1}{\beta r} t^2,$$

the differential expression (η) becomes

$$\sqrt{\left(\frac{2\beta}{r}\right)} dt \,.\, e^{\frac{r\beta}{2} \cdot \frac{\cos^2 z}{\sin^2 z} - t^2},$$

and consequently if we put

$$\int e^{-t^2} dt = e^{-\frac{\beta r}{2} \cot^2 z} \psi(r) \text{.................} (\text{A}),$$

where the integral is to be taken between the limits

$$t = \cot z \sqrt{\left(\frac{\beta r}{2}\right)}, \text{ and } t = \infty,$$

we get

$$\int_0^\infty \frac{\beta ds' \sin z \, e^{-r\beta s'}}{(\cos^2 z + 2s' \sin^2 z)^{\frac{1}{2}}} = \sqrt{(2\beta)} \frac{\psi(r)}{\sqrt{r}};$$

therefore if the refraction be represented by δz,

$$\delta z = \frac{\alpha}{1 - \alpha} \sqrt{(2\beta)} \left\{ \begin{aligned} & \psi(1) \\ & + \frac{\alpha\beta}{\sin^2 z} \{2^{\frac{1}{2}} \psi(2) - \psi(1)\} \\ & + \frac{\alpha^2\beta^2}{1.2 \sin^4 z} \{3^{\frac{3}{2}} \psi(3) - 2.2^{\frac{3}{2}} \psi(2) + \psi(1)\} \\ & + \frac{\alpha^3\beta^3}{1.2.3 \sin^6 z} \{4^{\frac{5}{2}} \psi(4) - 3.3^{\frac{5}{2}} \psi(3) + 3.2^{\frac{5}{2}} \psi(2) - \psi(1)\} \\ & + \&\text{c.} \text{ ....} \end{aligned} \right.$$

an expression for which, since

$$1 - x + \frac{x^2}{1\,.\,2} - \frac{x^3}{1\,.\,2\,.\,3} + \ldots = e^{-x},$$

we can also write the following:

$$\delta z = \frac{\alpha}{1-\alpha}\sqrt{(2\beta)}\left\{\begin{array}{l} e^{-\frac{\alpha\beta}{\sin^2 z}}\psi(1) \\ + \dfrac{\alpha\beta}{\sin^2 z}2^{\frac{1}{2}}e^{-\frac{2\alpha\beta}{\sin^2 z}}\psi(2) \\ + \dfrac{\alpha^2\beta^2}{1\,.\,2\,.\sin^4 z}3^{\frac{3}{2}}e^{-\frac{3\alpha\beta}{\sin^2 z}}\psi(3) \\ + \&\text{c.}\ldots\ldots\ldots\ldots\ldots\ldots \end{array}\right\}\ldots\ldots(\text{B}).$$

The calculation of the amount of refraction is thus made to depend upon the calculation of the transcendent $\psi(r)$, or upon $\int e^{-t^2}dt$, taken between the limits of $t = \sqrt{\left(\frac{\beta r}{2}\right)}\,.\cot z$, and $t = \infty$. Knowing then this and the numerical values of the constants α and β, we can find, by the formula (B), the amount of refraction δz for every apparent zenith-distance z.

8. Putting

$$\frac{r\beta}{2}\cot^2 z = T^2,$$

we have now therefore to determine the transcendent

$$\psi(r) = e^{T^2}\int_T^\infty e^{-t^2}dt.$$

For calculating this quantity two methods are preferred. The first expands the transcendent in a series, which is obtained by integration by parts, and which is continued to infinity, but from which we can nevertheless obtain the value to any required accuracy, since it possesses the property that if we stop at any particular term, the following terms do not together amount to more than the one last taken. We have, viz.:

$$\int e^{-t^2}dt = \int \frac{d\left(-\frac{1}{2}e^{-t^2}\right)}{dt}\,.\,\frac{dt}{t},$$

or, integrating by parts,

$$= -\frac{1}{2}\frac{e^{-t^2}}{t} - \frac{1}{2}\int e^{-t^2}\frac{dt}{t^2};$$

similarly we obtain

$$-\frac{1}{2}\int e^{-t^2}\frac{dt}{t^2} = -\frac{1}{2}\int \frac{d\left(-\frac{1}{2}e^{-t^2}\right)}{dt} \cdot \frac{dt}{t^3}$$

$$= +\frac{1}{2}\cdot\frac{1}{2}\cdot\frac{e^{-t^2}}{t^3} + \frac{1}{2}\cdot\frac{3}{2}\int e^{-t^2}\frac{dt}{t^4},$$

$$\frac{3}{4}\int e^{-t^2}\frac{dt}{t^4} = \frac{3}{4}\int \frac{d\left(-\frac{1}{2}e^{-t^2}\right)}{dt}\frac{dt}{t^5} = -\frac{3}{4}\cdot\frac{1}{2}\frac{e^{-t^2}}{t^5} - \&c....$$

consequently, at last

$$\int e^{-t^2}dt = -\frac{e^{-t^2}}{2t}\left\{1 - \frac{1}{2t^2} + \frac{1.3}{(2t^2)^2} - \frac{1.3.5}{(2t^2)^3} + \ldots\right.$$

$$\left. + \frac{1.3.5\ldots(2n-1)}{(2z^2)^n} \mp \frac{1.3.5\ldots(2n+1)}{2^{n+1}}\int e^{-t^2}\frac{dt}{t^{2n+2}}\right\},$$

or, putting in the limits,

$$\int_T^\infty e^{-t^2}dt = \frac{e^{-T^2}}{2T}\left\{1 - \frac{1}{2T^2} + \frac{1.3}{(2T^2)^2} - \frac{1.3.5}{(2T^2)^3} + \ldots\right.$$

$$\left. + \frac{1.3.5\ldots(2n-1)}{(2T^2)^n} \mp \frac{1.3.5\ldots(2n+1)}{2^{n+1}}\int_T^\infty e^{-t^2}\frac{dt}{t^{2n+2}}\right\}.$$

The factors of the numerator constantly increase; they will therefore at length become greater than $2T^2$, and from this point then every term increases continually, since the quantity put into the numerator is greater than the corresponding quantity put into the denominator.

But considering now the remainder

$$\mp \frac{1.3.5\ldots(2n+1)}{2^{n+1}}\int_T^\infty e^{-t^2}\frac{dt}{t^{2n+2}},$$

it will be easy to shew that this is smaller than the term immediately preceding. The value of the integral for instance is evidently smaller than

$$\int_T^\infty \frac{dt}{t^{2n+2}},$$

multiplied into the greatest value of e^{-t^2} between the limits T and ∞, that is to e^{-T^2}; and since now

$$\int_T^\infty \frac{dt}{t^{2n+2}} = \frac{1}{2n+1} \cdot \frac{1}{T^{2n+1}}$$

the remainder will always be smaller than

$$\mp \frac{1.3.5\ldots(2n-1)}{2^{n+1}T^{2n+1}} e^{-T^2}.$$

But this expression is nothing more than the previous term with the contrary sign. Consequently, if for example we leave off with a negative term, the remainder will be positive, but smaller than the previous term. Therefore, in order to obtain the correctest value possible of the transcendent by the computation of the series, we require only to continue it up to a term which is very small, and there is then to be feared only an error which is smaller than that previous very small term.

The second method of calculation consists in transforming, as Laplace was the first to show how, the transcendent into a continued fraction. Putting

$$e^{t^2}\int_T^\infty e^{-t^2}dt = U \ldots\ldots\ldots\ldots\ldots\ldots\ldots\ldots(\alpha),$$

we have

$$\frac{dU}{dt} = 2te^{t^2}\int_T^\infty e^{-t^2}\,dt + e^{t^2}e^{-t^2}$$

$$= 2tU + 1 \ldots\ldots\ldots\ldots\ldots\ldots\ldots\ldots\ldots\ldots(\beta).$$

But the nth differential coefficient of a product xy is

$$\frac{d^n.xy}{dt^n} = \frac{d^n x}{dt^n}y + n\frac{d^{n-1}x}{dt^{n-1}}\frac{dy}{dt} + \frac{n(n-1)}{1.2}\frac{d^{n-2}x}{dt^{n-2}}\frac{d^2y}{dt^2} + \&c.\ldots;$$

therefore we have also

$$\frac{d^{n+1}U}{dt^{n+1}} = 2t\frac{d^nU}{dt^n} + 2n\frac{d^{n-1}U}{dt^{n-1}};$$

an equation which can be written in the following manner, if we represent the product $1.2.3\ldots n$ by $n!$:

$$\frac{n+1}{(n+1)!}\frac{d^{n+1}U}{dt^{n+1}} = 2t\frac{d^n U}{n!\,dt^n} + 2\frac{d^{n-1}U}{(n-1)!\,dt^{n-1}},$$

or, if $\frac{d^n U}{n!\,dt^n}$ be represented by U_n,

$$(n+1)\,U_{n+1} = 2t\,U_n + 2\,U_{n-1}.$$

This equation is to be used from $n=1$, in which case U_0 becomes U. We obtain from it

$$-2\frac{U_{n-1}}{U_n} = 2t - (n+1)\frac{U_{n+1}}{U_n};$$

consequently

$$-\frac{1}{2}\frac{U_n}{U_{n-1}} = \frac{1}{2t-(n+1)\dfrac{U_{n+1}}{U_n}} = \frac{\dfrac{1}{2t}}{1-(n+1)\dfrac{U_{n+1}}{2t\,U_n}},$$

or

$$-\frac{U_n}{2t\,U_{n-1}} = \frac{\dfrac{1}{2t^2}}{1-(n+1)\dfrac{U_{n+1}}{2t\,U_n}} \quad \ldots\ldots\ldots\ldots\ldots (\gamma).$$

But, referring to equation (β),

$$\frac{U_1}{U} = 2t + \frac{1}{U};$$

therefore

$$U = \frac{-1}{2t-\dfrac{U_1}{U}} = \frac{-\dfrac{1}{2t}}{1-\dfrac{1}{2t}\dfrac{U_1}{U}};$$

but it follows from equation (γ), that

$$-\frac{1}{2t}\frac{U_1}{U} = \frac{\dfrac{1}{2t^2}}{1-2\dfrac{1}{2t}\dfrac{U_2}{U_1}};$$

substituting then this in the former equation and continuing the development, we obtain

$$U = -\cfrac{\frac{1}{2t}}{1 + \cfrac{\frac{1}{2t^2}}{1 + 2\cfrac{\frac{1}{2t^2}}{1 + 3\cfrac{\frac{1}{2t^2}}{1 + \&c.;}}}}$$

and, if we put $\frac{1}{2T^2} = q$,

$$2Te^{T^2}\int_T^\infty e^{-t^2}\,dt = \cfrac{1}{1 + \cfrac{q}{1 + \cfrac{2q}{1 + \cfrac{3q}{1 + \cfrac{4q}{1 + \&c.}}}}} \qquad (a)$$

If t is very small, we are enabled advantageously to employ a third method for calculating the transcendent. We have, viz.:

$$\int_T^\infty e^{-t^2}\,dt = \int_0^\infty e^{-t^2}\,dt - \int_0^T e^{-t^2}\,dt.$$

But the value of this last integral is easily obtained by expanding e^{-t^2} in a series, viz.:

$$\int_0^T e^{-t^2}\,dt = T - \frac{T^3}{3} + \frac{1}{2}\frac{T^5}{5} - \frac{1}{2\,.\,3}\cdot\frac{T^7}{7} + \ldots\ldots$$

and since

$$\int_0^\infty e^{-t^2}\,dt = \frac{\sqrt{\pi}}{2},$$

the value of

$$\int_T^\infty e^{-t^2}\,dt$$

can be obtained from it, which must further be multiplied by e^{T^2} in order to find the value of the transcendent represented by ψ.

We are now always enabled, from what has been shewn, to calculate the value of $\psi(r)$. On account of the constant employment of this transcendent it has been brought into a tabular form, and may be found, for example, in Bessel's *Fundamenta Astronomiæ.* The first division of the table thus given has T for the argument, and gives the values for every hundredth part from $T=0$ to $T=1$. But since the transcendent is more nearly inversely proportional to its argument the greater T is, Bessel chooses for the values which have a greater argument than $T=1$, the common logarithms of T as arguments. This second division of the table then extends from the logarithm 0·00 to the logarithm 1·00, or from $T=1$ to $T=10$, which will suffice for every requirement.

9. The formula (B) contains the two constants α and β, whose numerical values must be known, if we wish to determine the amount of refraction for any zenith-distance z. If the state of the atmosphere were always the same, these constants would also always have the same values. But since the density of the atmosphere depends upon the readings of the barometer and thermometer at the time of observation, α, which indeed is no other than the quantity $\frac{\mu^2_0 - 1}{2\mu^2_0}$ (where μ_0 represents the index of refraction from vacuum into the stratum of air at the surface of the earth*), will likewise be a function of the same. Similarly β or $\frac{h-l}{hl}\cdot\alpha$ will depend upon the height of the thermometer, since the quantity l or the height of a column of air of the density ρ_0, which corresponds to the pressure of the atmosphere, is a function of the temperature of the air. Consequently in order now to find the refraction for a given state of the atmosphere, that is, for given readings of the barometer and thermometer, we must in the calculation of the formula (B) employ the values of the constants really corresponding to this state of the atmosphere.

Now the atmosphere from 0^0 to 100^0 of the Centigrade Ther-

* Or also $\frac{c\rho_0}{2(1+c\rho_0)}$.

mometer is expanded by $\frac{3}{8}$ of its volume or more correctly by 0·36438. Consequently, if a volume of air of a given temperature, which Bessel has taken 8° Réaumur = 10° Celsius = 50° Fahrenheit*, be equal to unity, the same volume of air at a temperature of τ^0 Fahrenheit is

$$1 + (\tau - 50)\,\frac{0{\cdot}36438}{180} = 1 + (\tau - 50) \times 0{\cdot}0020243.$$

If therefore l_0 represents the value of the constant l for the temperature $\tau = 50^\circ$, then for every other temperature τ,

$$l = l_0\,\{1 + (\tau - 50)\,0{\cdot}0020243\},$$

and similarly, if ρ_0 is the density of the air for $\tau = 50^\circ$, for any other temperature

$$\rho = \frac{\rho_0}{1 + (\tau - 50)\,0{\cdot}0020243}.$$

But the density of the air depends also upon the height of the barometer; and since it also, according to Mariott's law, varies directly as the pressure, we shall have, if we again represent by ρ_0 the density for a given height of the barometer, which Bessel takes at 29·6 English inches, then for any other barometer-reading b,

$$\rho = \rho_0 \cdot \frac{b}{29{\cdot}6}.$$

Since now α, for changes of the density so small as here come under consideration, can be assumed to be proportional to this density, we obtain, if we call the value of α for the normal condition of the atmosphere, α_0,

$$\alpha = \frac{\alpha_0 \frac{b}{29{\cdot}6}}{1 + (\tau - 50)\,0{\cdot}0020243}.$$

Now Bessel, in his *Fundamenta Astronomiæ*, has determined the value of the constant α_0 from Bradley's observations, and found from thence, that

$$\alpha_0 = 57'',538,$$

* According to the thermometer employed by Bradley.

and further for the constant h,

$$h = 116865{\cdot}8 \text{ toises.}$$

From this follows, if we assume the radius of curvature for the Greenwich Observatory to be equal to 3269805 toises, the value of the constant β for the normal condition of the atmosphere,

$$\text{namely } \beta_0 = 745{\cdot}747,$$

with which we have all the required data, in order to be able to calculate, from the formula (B), the refraction for any zenith-distance and any condition of the atmosphere. If for example there were required the refraction for 80° zenith-distance and for the normal condition of the atmosphere, we have

$$\log \frac{\alpha}{1-\alpha} \sqrt{(2\beta)} = 3{\cdot}34688.$$

Moreover, we obtain for the logarithms of the separate functions ψ, according to the formulæ (a) or (b) in No. 8, and for the logarithms of the factors of these quantities, the following values:

	$\log \{n^{\frac{2n-3}{2}}\}$.	$\log \left\{\frac{1}{n-1!} \left(\frac{\alpha\beta}{\sin^2 z}\right)^{n-1}\right\}$.	$\log \psi(n)$.	$\log \left\{e^{-n} \cdot \frac{\alpha\beta}{\sin^2 z}\right\}$.
$n = 1$,	0·00000	0·00000	9·14982	9·90685
$n = 2$,	0·15051	9·33142	9·00745	9·81369
$n = 3$,	0·71568	8·36181	8·92228	9·72054
$n = 4$,	1·50515	7·21610	8·86128	9·62738
$n = 5$,	2·44640	5·94546	8·81362	9·53423
$n = 6$,	3·46568	4·57791	8·77473	9·44108
$n = 7$,	4·64804	3·13118	8·74169	9·34792

From these we obtain then for the separate terms of the formula (B) within the brackets,

first term = 0·113938,
second ...... = 0·020094,
third ...... = 0·005252,
fourth ...... = 0·001622,
fifth ...... = 0·000549,

sixth term = 0·000182
seventh ...... = 0·000074
sum = 0·141711
log = 9· 15140
log const. = 3· 34688
refraction = + 5′.15″,0

Bessel has more recently found that the refractions calculated in this manner must be multiplied by 1·003282, in order to represent the Königsberg observations. Performing this multiplication, we obtain for the refraction at 80° zenith-distance and for the normal condition of the atmosphere

$$+ 5'.16'',0.$$

If the refraction were required not for the normal condition of the atmosphere but for a temperature of τ^0 Fahrenheit and for a height of b inches of the barometer, we must first seek the values of α and β corresponding to these according to the formula given above, and with these perform the calculation of (B).

But in order to do away with the necessity of this troublesome calculation, the refractions have been reduced to tables which have the apparent zenith-distance for the argument. One table gives the values of the refraction for the normal temperature and the normal barometer height, or of the so-called *mean refraction.* Another table then gives the corrections, which must be added to this mean refraction, in order to obtain the true refractions corresponding to the existing conditions of the atmosphere. For the calculation of this last table it is necessary to investigate the analytical expressions for the changes in the refraction dependent upon the readings of the thermometer and barometer.

10. Let r denote the true and δz the main refraction, then,

$$r = \delta z + \frac{d\delta z}{d\tau}(\tau - 50) + \frac{d\delta z}{db}(b - 29{\cdot}6) \ldots\ldots\ldots\ldots (a).$$

But, by formula (B) in No. 8 we have by putting

$$\frac{\alpha\beta}{\sin^2 z} = x,$$

$$(1-\alpha)\,\delta z = \sin^2 z\sqrt{\frac{2}{\beta}}\,.\,x\Sigma\left\{n^{\frac{2n-3}{2}}\frac{x^{n-1}}{1\,.\,2\,.\,3\,\ldots\,(n-1)}\,e^{-nx}\,.\,\psi(n)\right\}$$

$$= \sin^2 z\sqrt{\frac{2}{\beta}}\,\Sigma\left\{n^{\frac{2n-1}{2}}\frac{x^n}{1\,.\,2\,.\,3\,\ldots\,n}\,e^{-nx}\,\psi(n)\right\},$$

where for n we must substitute all whole numbers beginning with unity. For the sake of brevity let

$$n^{\frac{2n-1}{2}}\psi(n) = q_n, \quad \text{and } \Sigma\frac{x^n\,.\,e^{-nx}}{1\,.\,2\,.\,3\,\ldots\,n}\,q_n = Q_n,$$

then the equation becomes simply

$$(1-\alpha)\,\delta z = \sin^2 z\sqrt{\frac{2}{\beta}}\,.\,Q_n.$$

On account of the smallness of α, we may consider the factor $1-\alpha$ as constant, so that the variables b and τ occur only in x and β, τ alone appearing in β, while in x both τ and b appear. Taking now first the differential coefficient of Q_n, we have

$$\frac{dQ_n}{d\tau} = \left(\frac{dQ_n}{dx}\right)\left(\frac{dx}{d\tau}\right) + \left(\frac{dQ_n}{d\beta}\right)\left(\frac{d\beta}{d\tau}\right) \ldots\ldots\ldots\ldots\ldots (b).$$

But we have also,

$$\left(\frac{dQ_n}{dx}\right) = \frac{1-x}{x}\,\Sigma\,\frac{x^n e^{-nx}}{1\,.\,2\,.\,3\,\ldots\,n}\,nq_n.$$

Or, if we denote the series $\Sigma\,\frac{x^n e^{-nx}}{1\,.\,2\,.\,3\,\ldots\,n}\,nq_n$ by Q_n',

$$\left(\frac{dQ_n}{dx}\right) = \frac{1-x}{x}\,Q_n' \ldots\ldots\ldots\ldots\ldots\ldots\ldots\ldots (c);$$

moreover

$$\left(\frac{dQ_n}{d\beta}\right) = \Sigma\,\frac{x^n\,.\,e^{-nx}}{1\,.\,2\,.\,3\,\ldots\,n}\left(\frac{dq_n}{d\beta}\right).$$

But we have also, by employing the known law for the calculation of the differential coefficient of a definite integral for one of its limits,

$$\left(\frac{dq_n}{d\beta}\right) = \frac{\cot^2 z}{2}\,nq_n - \frac{\cot z\sqrt{\frac{\beta}{2}}}{2\beta}\,n^n;$$

therefore we shall also have

$$\left(\frac{dQ_n}{d\beta}\right) - \frac{\cot^2 z}{2} Q'_n = \frac{\cot z}{2\sqrt{(2\beta)}} \Sigma \frac{n^n x^n e^{-nx}}{1.2.3 \ldots n} \ldots\ldots\ldots (d).$$

Moreover, since

$$\left(\frac{d\beta}{dl}\right) = -\frac{\alpha}{l^2}, \quad \left(\frac{d\alpha}{d\tau}\right) = -\alpha\epsilon, \quad \left(\frac{dl}{d\tau}\right) = l\epsilon^*,$$

$$\left(\frac{d\beta}{d\tau}\right) = -\epsilon\beta \frac{h}{h-l},$$

$$\left(\frac{dx}{d\tau}\right) = \frac{x}{\beta}\left(\frac{d\beta}{d\tau}\right) + \frac{x}{\alpha}\left(\frac{d\alpha}{d\tau}\right) = -\epsilon x \left(\frac{2h-l}{h-l}\right) \ldots\ldots\ldots (e).$$

Consequently from the formulæ (*b*), (*c*), (*d*), (*e*),

$$\left(\frac{dQ_n}{d\tau}\right) = -\epsilon Q'_n \left\{\frac{2h-l}{h-l}(1-x) + \frac{h}{h-l} \cdot \frac{\beta}{2} \cot^2 z\right\}$$

$$+ \epsilon \sqrt{\frac{\beta}{2}} \cdot \frac{\cot z}{2} \left(\frac{h}{h-l}\right) \Sigma \frac{n^n . x^n . e^{-nx}}{1.2.3 \ldots n} \ldots\ldots\ldots\ldots (f).$$

Since moreover the variable *b* only appears in *x*, we have

$$\frac{dQ_n}{db} = \left(\frac{dQ_n}{dx}\right)\left(\frac{dx}{db}\right) = \frac{1-x}{x} Q'_n \left(\frac{dx}{db}\right),$$

or since

$$\left(\frac{dx}{db}\right) = \frac{x}{\alpha}\left(\frac{d\alpha}{db}\right),$$

$$\left(\frac{dQ_n}{db}\right) = \frac{1-x}{29 \cdot 6} \cdot Q'_n \ldots\ldots\ldots\ldots\ldots\ldots\ldots\ldots (g).$$

But, by differentiating the expression for $(1-\alpha)\,\delta z$, we obtain

$$(1-\alpha)\frac{d.\delta z}{d\tau} = -\frac{1}{2}\delta z . \frac{1}{\beta}\left(\frac{d\beta}{d\tau}\right) + \sin^2 z . \sqrt{\frac{2}{\beta}} \cdot \left(\frac{dQ_n}{d\tau}\right),$$

and

$$(1-\alpha)\frac{d.\delta z}{db} = +\sin^2 z \sqrt{\frac{2}{\beta}} \cdot \left(\frac{dQ_n}{db}\right).$$

* By the Formula in No. 9, where ϵ represents the number 0·0020243.

Substituting in these the values of the differential coefficients of Q_n from the equations (f) and (g), and, in the same manner, for $\left(\frac{d\beta}{d\tau}\right)$ its value from the first of equations (e), we find at length,

$$(1-\alpha)\frac{d\,.\,\delta z}{d\tau} = \epsilon\,\delta z\left(\frac{\frac{1}{2}h}{h-l}\right) - \epsilon\,Q'_n \sin^2 z\sqrt{\frac{2}{\beta}}\left\{\frac{2h-l}{h-l}(1-x)\right.$$

$$\left.+\left(\frac{h}{h-l}\right)\frac{\beta}{2}\cot^2 z\right\} + \epsilon\sin^2 z\cot z\left(\frac{\frac{1}{2}h}{h-l}\right)\Sigma\,\frac{n^n\,.\,x^n\,.\,e^{-nx}}{1\,.\,2\,.\,3\,\ldots\,n},$$

.... (A),

and $$(1-\alpha)\frac{d\,.\,\delta z}{db} = \sin^2 z\sqrt{\frac{2}{\beta}}\,.\,\frac{1-x}{29{\cdot}6}\,Q'_n.$$

By means of these formulæ we could now reduce to a tabular form both the values of the differential coefficients and the values of the mean refraction, the argument being the apparent zenith-distance z, and then compute the true refraction by means of the formula (a). This formula is however not convenient. For more convenient logarithmic computation, make

$$r = \frac{\delta z}{\{1+\epsilon(\tau-50)\}^{\lambda}}\left(\frac{b}{29{\cdot}6}\right)^{A} \quad \ldots\ldots\ldots\ldots\ldots\ldots (B),$$

then are λ and A functions of $\frac{d\,.\,\delta z}{d\tau}$ and $\frac{d\,.\,\delta z}{db}$, which can themselves also be tabulated.

These functions can be now easily calculated. Since, namely,

$$\{1+\epsilon(\tau-50)\}^{-\lambda} = 1-\lambda\epsilon(\tau-50),\ \text{\&c.},$$

$$\left(\frac{b}{29{\cdot}6}\right)^{A} = 1 + A\left(\frac{b}{29{\cdot}6}-1\right),\ \text{\&c.},$$

we obtain from equation (B), taking only the first terms of the series,

$$r = \delta z - \lambda\epsilon(\tau-50)\,\delta z + A\left(\frac{b}{29{\cdot}6}-1\right)\delta z,$$

and, by comparing this equation with the formula (a),

$$\left.\begin{aligned} \lambda &= -\frac{1}{0{\cdot}0020243\,\delta z}\cdot\frac{d\,.\,\delta z}{d\tau} \\ A &= \frac{29{\cdot}6}{dz}\cdot\frac{d\,.\,\delta z}{db} \end{aligned}\right\} \quad\ldots\ldots\ldots\ldots\ldots\ldots (C).$$

Example.

For apparent zenith-distance 80° we obtain, by taking into consideration the terms which contain $\psi(8)$ of which the logarithm is $8{\cdot}71302$,

$$\log Q'_n = 8{\cdot}58950,$$

$$\log \Sigma \frac{n^n x^n e^{-nx}}{1\,.\,2\,.\,3 \ldots n} = 9{\cdot}43611,$$

and, at the same time,

$$\log\left(\frac{1}{\epsilon}\cdot\frac{d\,\delta z}{d\tau}\right) = -\,7{\cdot}20207,$$

$$\log\frac{d\,\delta z}{db} = 5{\cdot}71441,$$

and lastly

$$\lambda = +\,1{\cdot}0428,$$

$$A = +\,1{\cdot}0042.$$

If, for example, we desired to calculate the refraction for $\tau = 15^\circ$ Réaumur and $b = 28{\cdot}6$ English inches, we should obtain, since 15° Réaumur $= 65^\circ{\cdot}7$ Fahrenheit,

$$\log\frac{1}{\{1 + 0{\cdot}0020243\,(\tau - 50)\}^\lambda} = 9{\cdot}98583,$$

$$\log\left(\frac{b}{29{\cdot}6}\right)^A = 9{\cdot}98502\,;$$

and hence, since

$$\delta z = +\,5'.\,16'',0,$$

$$r = +\,4'.\,55'',5.$$

Bessel's Refraction Tables are constructed according to formulæ (B) and (C). The first table gives, with zenith-distance for argument, besides the mean refraction, the quantities A and λ. The other tables give, with arguments the temperature observed according to any one of the three scales in use, and the barometer-reading either in Paris inches, English inches, or metres, the logarithms of the factors

$$\frac{1}{\{1+0{\cdot}0020243\,(\tau-50)\}} \text{ and } \frac{b}{29{\cdot}6}.$$

The first factor is the ratio of a volume of air at the temperature $\tau = 50^0$ of the Fahrenheit thermometer used by Bradley, to the volume at another temperature. If we denote by 1 the volume of air at the freezing point, then will the volume at any other temperature be

$$1+0{\cdot}0020243\,(\tau-32^0)$$
$$=1+\frac{0{\cdot}36438}{180}(\tau-32^0).$$

Now Bessel has found that the thermometer employed by Bradley gave all temperatures too high by $1^0{\cdot}25$, while the freezing point of the same was too low by the same quantity. Thus the temperature 50^0 corresponded to true temperature $48^0{\cdot}75$. And thus, if we denote by γ the coefficient

$$\frac{1}{1+0{\cdot}0020243\,(\tau-50)}$$

$$\gamma=\frac{180+16{\cdot}75\times 0{\cdot}36438}{180+(f-32)\,0{\cdot}36438} \quad\ldots\ldots\ldots\ldots\ldots (D),$$

where f is the temperature of the air expressed according to the scale of a Fahrenheit thermometer. If we denote by r and c the same temperature according to Réaumur and Celsius, we have

$$\gamma=\frac{180+16{\cdot}75\times 0{\cdot}36438}{180+\frac{9}{4}r\times 0{\cdot}36438}$$

$$=\frac{180+16{\cdot}75\times 0{\cdot}36438}{180+\frac{9}{5}c\times 0{\cdot}36438}.$$

By means of these formulæ $\log\gamma$ has been tabulated.

For the barometer the reading 29·6 English inches was taken as the normal reading. Since now Bessel has found that this (Bradley's) instrument gave all barometer-heights too small by half of a Paris line, this normal reading becomes 29·644 English inches or 333·78 Paris lines. Barometers are now always divided either according to Paris lines, or English inches, or metres. The lengths of a Paris line, an English inch, and the metre correspond to the normal temperature 13° Réaumur, 62° Fahrenheit, and 0° Celsius. Denote then by $b^{(l)}$, $b^{(e)}$, and $b^{(m)}$ the height of the barometer expressed in terms of a Paris line, an English inch, and the metre, as they are observed at any temperature t, then these will not serve as the true measure, but, if s denote the extension of the scale from the freezing to the boiling point, then will the barometer-height read at the temperature t be to that which would have been read if t had been equal to the normal temperature T, in the proportion of

$$1+\frac{s}{a}T : 1+\frac{s}{a}t,$$

if the length of the scale at the freezing point be taken for the unit, and a denote the number of divisions between the freezing and boiling points of the thermometer. Denoting then again by r, f, c the observed temperatures according to Réaumur, Fahrenheit, and Celsius, the height of the barometer referred to the true standard will be

$$b^{(l)}.\frac{80+rs}{80+13s}, \quad b^{(e)}.\frac{180+(f-32)s}{180+30s}, \quad b^{(m)}.\frac{100+cs}{100},$$

where $s=0·0018782$ on a scale of brass.

Since now an English inch $=\frac{1·065765}{12}$ Paris lines, and a metre $=443·296$ Paris lines, the three preceding barometer-heights are, in Paris lines,

$$b^{(l)}.\frac{80+rs}{80+13s}=b^{(e)}.\frac{12}{1·065765}.\frac{180+(f-32)s}{180+30s}$$

$$=b^{(m)}.443·296.\frac{100+cs}{100} \dots\dots\dots\dots\dots (a).$$

The normal height of the barometer above exhibited or 333·78 Paris lines corresponds to the normal temperature 8° Réaumur, or 50° Fahrenheit, or 10° Celsius, and is therefore also measured on a scale of these temperatures. The normal barometer-height reduced to true Paris measure will therefore be

$$B_0 = 333{\cdot}78 \, . \frac{80 + 8s}{80 + 13s},$$

and by this quantity are the observed barometer-heights in (α), when reduced to true Paris measure, to be divided.

Still however allowance must be made for the expansion of the mercury, which, from the freezing point to the boiling point, is equal to $\frac{1}{55{\cdot}5}$ part. Denoting this number by q, the barometer-height observed at a temperature t will be to that which would have been observed, if t had been equal to the normal temperature T, as

$$1 + \frac{q}{\alpha} t \; : \; 1 + \frac{q}{\alpha} T.$$

We obtain therefore for the three different thermometers the following correction-factors, by which the barometer-heights in (α) are to be multiplied:

$$\frac{80 + 8q}{80 + rq}, \quad \frac{180 + 18q}{180 + (f - 32)\,q}, \quad \text{and} \quad \frac{100 + 10q}{100 + cq},$$

where r, f, and c are the readings of the thermometer attached to the barometer. The complete expression for $\frac{b}{29{\cdot}6}$ will therefore consist of two factors, of which one depends solely on the height of the barometer, the other solely on the temperature of the barometer. Denoting the first by B, the other by T, we have

$$\left.\begin{aligned} B &= \frac{b^{(l)}}{333{\cdot}78} \cdot \frac{80 + 8q}{80 + 8s} \\ &= \frac{b^{(e)}}{333{\cdot}78} \cdot \frac{12}{1{\cdot}065765} \cdot \frac{80 + 13s}{80 + 8s} \cdot \frac{180 + 18q}{180 + 30s} \\ &= \frac{b^{(m)}}{333{\cdot}78} \cdot 443{\cdot}296 \cdot \frac{80 + 13s}{80 + 8s} \cdot \frac{100 + 10q}{100} \\ \text{and} \quad & \\ T &= \frac{80 + rs}{80 + rq} = \frac{180 + (f - 32)s}{180 + (f - 32)q} = \frac{100 + cs}{100 + cq} \end{aligned}\right\} \quad \cdots\cdots (E).$$

By these formulæ (E) and (D) are the tables constructed which give $\log B$ with argument the barometer-height according to the three scales, and $\log T$ with argument the height of the thermometer attached to the barometer (the interior thermometer) according to the three thermometer scales, and finally $\log \gamma$ with argument the height of the thermometer suspended in free air (the exterior thermometer) in like manner for all three scales.

These refraction tables of Bessel are to be found in Bessel's *Tabulæ Regiomontanæ*, in Schumacher's *Hülfstafeln*, and also in the *Astronomischen Jahrbuchern* of Encke. Instead of the quantity δz Bessel gives the quantity α, computed by the equation

$$\delta z = \alpha \tan z,$$

so that the expression for the refraction is the following:

$$\log r = \log \alpha + \log \tan z + \lambda \log \gamma + A\,(\log B + \log T) \ldots (F).$$

The fundamental constant for the computation of α is $57'',538$ multiplied by $1{\cdot}003282$ or $57''{\cdot}727$.

11. The theory of refraction developed in the preceding pages is that given by La Place and Bessel. It corresponds to observations very completely to the greatest possible zenith-distances. There are however still other formulæ for refraction which are based on simpler laws for the density of the air, and which therefore give much simpler expressions for the refraction, but which deviate from observations considerably for great zenith-distances. Since however it frequently happens that the simpler analytical expressions are convenient in practice, it is desirable in the following pages to deduce the most important of them.

In No. 6 was investigated the differential equation (g)

$$d\zeta' = -\frac{\alpha \dfrac{d\rho}{\rho_0}(1-s)\sin z}{\left\{1 - 2\alpha\left(1 - \dfrac{\rho}{\rho_0}\right)\right\}\sqrt{\left\{\cos^2 z - 2\alpha\left(1 - \dfrac{\rho}{\rho_0}\right) + (2s - s^2)\sin^2 z\right\}}}.$$

This equation is very easily integrated, if between s and r we assume the law

$$1 - s = \left\{1 - 2\alpha\left(1 - \frac{\rho}{\rho_0}\right)\right\}^m,$$

where m is an unknown quantity to be determined by observation. The equation will then for instance become

$$d\zeta' = -\frac{\alpha \sin z \,.\, \frac{d\rho}{\rho_0}\left\{1 - 2\alpha\left(1 - \frac{\rho}{\rho_0}\right)\right\}^{\frac{2m-3}{2}}}{\sqrt{\left[1 - \left\{1 - 2\alpha\left(1 - \frac{\rho}{\rho_0}\right)\right\}^{2m-1}\sin^2 z\right]}};$$

or, by introducing another variable, given by the equation

$$\left\{1 - 2\alpha\left(1 - \frac{\rho}{\rho_0}\right)\right\}^{\frac{2m-1}{2}} \sin z = w,$$

very simply

$$d\zeta' = \frac{dw}{(2m-1)\sqrt{(1-w^2)}},$$

and so by integration

$$\delta z = \frac{1}{2m-1} \sin^{-1} w.$$

Since now at the surface of the earth the density of the air is equal to ρ_0, but is equal to 0 at the boundary of the atmosphere, we must take for the limits of the integral

$$w = \sin z \text{ and } w = (1-2\alpha)^{\frac{2m-1}{2}} \sin z,$$

and we then obtain

$$\delta z = \frac{1}{2m-1}\left\{z - \sin^{-1}\left[(1-2\alpha)^{\frac{2m-1}{2}} \sin z\right]\right\},$$

or

$$\sin\left[z - (2m-1)\,\delta z\right] = (1-2\alpha)^{\frac{2m-1}{2}} \sin z,$$

for which may be written briefly

$$M \sin z = \sin(z - N\delta z) \quad \ldots\ldots\ldots\ldots\ldots\ldots \quad (a).$$

If $z = 90°$, we have, denoting the corresponding value of δz, that is the horizontal refraction, by h,

$$M = \cos Nh,$$

and thus we have in general

$$\cos Nh \sin z = \sin(z - N\delta z) \quad \ldots\ldots\ldots\ldots\ldots \quad (b).$$

This is the rule for refraction given by Simpson, which however was not investigated by him analytically but by practical means. If the coefficient be suitably determined, the refraction can be by means of it perfectly well determined to 85° zenith-distance.

By adding to equation (a) the identical equation

$$\sin z = \sin z,$$

we obtain

$$\sin z\,\{1 + M\} = 2 \sin\left(z - \frac{N}{2}\,\delta z\right)\cos\frac{N}{2}\,\delta z.$$

And again, by subtracting the same identical equation, we find

$$\sin z\,\{1 - M\} = 2 \cos\left(z - \frac{N}{2}\,\delta z\right)\sin\frac{N}{2}\,\delta z.$$

Dividing one equation by the other, we obtain

$$\tan\frac{N}{2}\,\delta z = \frac{1-M}{1+M}\tan\left(z - \frac{N}{2}\,\delta z\right),$$

or, by again introducing the horizontal refraction,

$$\tan\frac{N}{2}\,\delta z = \tan\left(\frac{N}{2}h\right)^2\tan\left(z - \frac{N}{2}\,\delta z\right) \;\ldots\ldots\ldots\ldots (c).$$

This is the rule for refraction proposed by Bradley, which can be briefly written

$$\tan\alpha\,\delta z = \beta\tan(z - \alpha\delta z).$$

If δz be small, we may be allowed to put

$$\delta z = \frac{\beta}{\alpha}\tan(z - \alpha\delta z),$$

from which it is seen that, as long as the refraction is small, it may be assumed to be proportional to the tangent of the apparent zenith-distance.

12. Since refraction causes all stars to appear at a greater altitude than that which they really have, it has also this

effect, that it renders objects visible when they are in fact beneath the horizon. It therefore accelerates the rising, and retards the setting of the heavenly bodies.

In general we have the equation

$$\sin h = \sin \phi \sin \delta + \cos \phi \cos \delta \cos t.$$

If now a star be seen in the horizon, it is really beneath it by a quantity which is equal to the horizontal refraction. Denoting this by ρ, we have thus for the rising and setting of a star the equation

$$-\sin \rho = \sin \phi \sin \delta + \cos \phi \cos \delta \cos t_0 \ldots\ldots\ldots (a).$$

Call now T the hour-angle which the star would have at its rising and setting if the refraction were equal to 0, and we shall thus have

$$0 = \sin \phi \sin \delta + \cos \phi \cos \delta \cos T \ldots\ldots\ldots\ldots\ldots (b).$$

Thus from the two equations (a) and (b)

$$\cos t_0 = \cos T \left(1 - \frac{\sin \rho}{\cos \phi \cos \delta \cos T}\right).$$

The expression is still simpler, if there be only required the correction ΔT of the hour-angle T which is produced by the refraction. This however is found with sufficient accuracy by differentiating the equation for $\sin h$ with respect to h and t; we then obtain, if ΔT be expressed in seconds of time,

$$\Delta T = \frac{\rho}{\cos \phi \,.\, \cos \delta \sin T} \cdot \frac{1}{15}.$$

For Arcturus and the latitude of Berlin there was found, in No. 12 of the first section,

$$T = 7^h.\ 53^m,7, = 118^\circ.\ 16',8.$$

Computing now the formula for ΔT on the assumption that $\rho = 33'$, we obtain

$$\Delta T = 4^m.\ 22^s,0.$$

By so much therefore will the rising or the setting of Arcturus at Berlin be accelerated or retarded.

The zenith-distance of a star culminating south of the zenith is $\phi - \delta$, and thus the south declination of a star, which at its upper culmination is exactly in the horizon, is $90^{\circ} - \phi$. (Section I. No. 12.) But since now such a star by virtue of the refraction appears at an altitude ρ, where ρ again denotes the horizontal refraction, we see that all stars pass above the south horizon whose south declination is less than $90^{\circ} - \phi + \rho$. In the same manner we see that all stars at their lower culminations appear above the north horizon whose north declination is greater than $90^{\circ} - \phi - \rho$.

REMARK. On Refraction compare Laplace, *Méc. Cél.* Livre x.; Bessel, *Fundamenta Astronomiæ*, page 26 et seq.; and the Preface to Bessel's *Tab. Regiomont.* page 59 et seq.

III. ABERRATION.

13. Since the velocity of the earth in its annual orbit round the sun has a fixed relation to the velocity of light, we see the stars from the surface of the earth thus moving, not in the direction in which they really are, but always in advance by a small angle in the direction in which the earth is moving. Imagine two separate instants of time t and t' at which the ray of light coming from a fixed star arrives successively at the object-glass and eye-piece of a telescope (or at the lenses and the retina of our eye). Let the positions of the object-glass and the eye-piece in space at the time t be a and b, and at the time t', a' and b'. Fig. 4. Then is the true direction of the ray of light in space the direction of the straight line ab', and on the contrary, the direction ab or $a'b'$, since the latter, on account of the infinite distance of the stars is parallel to ab, is the direction of the apparent place which is observed. The difference between the direction $b'a$ and ba is called the *annual aberration* of the fixed stars.

Let now x, y, z be the rectangular co-ordinates of the eye-piece b at the time t referred to any fixed point in space; then will

$$x + \frac{dx}{dt}(t' - t),\ y + \frac{dy}{dt}(t' - t), \text{ and } z + \frac{dz}{dt}(z' - z)$$

be the co-ordinates of the eye-piece at the time t', since we may consider the motion of the earth to be linear during the short interval $t' - t$. Let the co-ordinates of the object-glass referred to the eye-piece be ξ, η, ζ; then are the co-ordinates of the object-glass at the time t at which the light enters it

$$x + \xi, \quad y + \eta, \quad z + \zeta.$$

Taking now the plane of the equator for the plane of xy, and the two other co-ordinate planes at right angles to it, so that the plane of xz passes through the equinoctial points, and that of yz through the solsticial points, denoting moreover the right ascension and declination of that point in which the true direction of the ray of light meets the visible sphere of the heavens by α and δ, and by μ the velocity of light, then will the latter in the time $t' - t$ describe a space whose projections on the three co-ordinate axes are

$$\mu\,(t' - t) \cos\delta \cos\alpha, \quad \mu\,(t' - t) \cos\delta \sin\alpha, \quad \mu\,(t' - t) \sin\delta.$$

In addition, calling the length of the telescope l, and the right ascension and declination of that point in which the apparent direction of the ray of light meets the sphere of the heavens, α' and δ'; the apparent co-ordinates of the object-glass referred to the eye-piece, which are observed, are

$$\xi = l \cos\delta' \cos\alpha', \quad \eta = l \cos\delta' \sin\alpha', \quad \zeta = l \sin\delta'.$$

The true direction of the ray of light is now given by means of the co-ordinates of the object-glass at the time t:

$$\begin{aligned} &l \cos\delta' \cos\alpha' + x, \\ &l \cos\delta' \sin\alpha' + y, \\ &l \sin\delta' \qquad\quad + z; \end{aligned}$$

and by the co-ordinates of the eye-piece at the time t',

$$x + \frac{dx}{dt}\,(t' - t),$$

$$y + \frac{dy}{dt}\,(t' - t),$$

$$z + \frac{dz}{dt}\,(t' - t).$$

We obtain therefore the following equations, denoting by L the quantity $\frac{l}{t'-t}$:

$$\mu \cos \delta \cos \alpha = L \cos \delta' \cos \alpha' - \frac{dx}{dt},$$

$$\mu \cos \delta \sin \alpha = L \cos \delta' \sin \alpha' - \frac{dy}{dt},$$

$$\mu \sin \delta \qquad = L \sin \delta' \qquad - \frac{dz}{dt}.$$

From these equations we can easily find

$$L \cos \delta' \cos (\alpha' - \alpha) = \cos \delta + \frac{1}{\mu}\left(\frac{dy}{dt} \sin \alpha + \frac{dx}{dt} \cos \alpha\right),$$

$$L \cos \delta' \sin (\alpha' - \alpha) = \qquad \frac{1}{\mu}\left(\frac{dy}{dt} \cos \alpha - \frac{dx}{dt} \sin \alpha\right),$$

or

$$\tan (\alpha' - \alpha) = \frac{\frac{1}{\mu} \sec \delta \left(\frac{dy}{dt} \cos \alpha - \frac{dx}{dt} \sin \alpha\right)}{1 + \frac{1}{\mu} \sec \delta \left(\frac{dy}{dt} \sin \alpha + \frac{dx}{dt} \cos \alpha\right)}.$$

We obtain a precisely similar equation for $\tan (\delta' - \delta)$. Developing both equations into series by the employment of formula 14 in No. 11 of the Introduction, we obtain, if we substitute in the formula for $\tan (\delta' - \delta)$ the value of $\tan \frac{1}{2} (\alpha' - \alpha)$ derivable from $\alpha' - \alpha$, to terms of the second order inclusive,

$$\left.\begin{aligned} \alpha' - \alpha = &- \frac{1}{\mu}\left(\frac{dx}{dt} \sin \alpha - \frac{dy}{dt} \cos \alpha\right) \sec \delta + \frac{1}{\mu^2}\left(\frac{dx}{dt} \sin \alpha - \frac{dy}{dt} \cos \alpha\right) \\ &\times \left(\frac{dx}{dt} \cos \alpha + \frac{dy}{dt} \sin \alpha\right) \sec^2 \delta + \&c. \\ \delta' - \delta = &- \frac{1}{\mu}\left(\frac{dx}{dt} \sin \delta \cos \alpha + \frac{dy}{dt} \sin \delta \sin \alpha - \frac{dz}{dt} \cos \delta\right) \end{aligned}\right\} \ldots (a)$$

$$- \frac{1}{2\mu^2}\left(\frac{dx}{dt} \sin \alpha - \frac{dy}{dt} \cos \alpha\right)^2 \tan \delta$$

$$+\frac{1}{\mu^2}\left(\frac{dx}{dt}\cos\delta\cos\alpha+\frac{dy}{dt}\cos\delta\sin\alpha+\frac{dz}{dt}\sin\delta\right)$$
$$\times\left(\frac{dx}{dt}\sin\delta\cos\alpha+\frac{dy}{dt}\sin\delta\sin\alpha-\frac{dz}{dt}\cos\delta\right).$$

Imagine now the place of the earth by means of co-ordinates x, y in the plane of the ecliptic to be referred to the centre of the sun, and take the line from the sun's centre towards the vernal equinox as the positive direction of the axis of x, and, as the positive direction of the axis of y, the perpendicular to it drawn to the summer solsticial colure, then, if we denote the longitude of the sun as seen from the earth by $\odot$, and its distance from the earth by R, we have*

$$x=-R\cos\odot,$$
$$y=-R\sin\odot.$$

If we refer the co-ordinates to the plane of the equator, taking as the axis of x the line drawn towards the vernal equinox and imagining the co-ordinate axis of z in the plane yz to be turned through an angle ϵ, equal to the obliquity of the ecliptic, we obtain

$$x=-R\cos\odot,$$
$$y=-R\sin\odot\cos\epsilon,$$
$$z=-R\sin\odot\sin\epsilon,$$

and from hence,

$$\left.\begin{aligned}\frac{dx}{dt}&=+R\sin\odot\frac{d\odot}{dt}\\ \frac{dy}{dt}&=-R\cos\odot\cos\epsilon\frac{d\odot}{dt}\\ \frac{dz}{dt}&=-R\cos\odot\sin\epsilon\frac{d\odot}{dt}\end{aligned}\right\}\ldots\ldots\ldots\ldots\ldots(b).$$

Substituting these values in equations (a) and introducing instead of μ the number of seconds of time required by light to traverse the radius of the earth's orbit, so that

$$\mu=\frac{R}{k},\ \text{and therefore}\ \frac{1}{\mu}=\frac{k}{R},$$

we have, retaining only the terms of the first order,

* Since the longitude of the earth as seen from the sun is $180^\circ+\odot$.

$$\alpha' - \alpha = -k\frac{d\odot}{dt}\left(\cos\odot\cos\epsilon\cos\alpha + \sin\odot\sin\alpha\right)\sec\delta$$

$$\delta' - \delta = +k\frac{d\odot}{dt}\{\cos\odot[\sin\alpha\sin\delta\cos\epsilon - \cos\delta\sin\epsilon] - \cos\alpha\sin\delta\sin\odot\}.$$

Now the sun moves* in a mean day 59′. 8″,33, so that

$$\frac{d\odot}{dt} = \frac{59'.\,8'',33}{86400} = 0{\cdot}0410686\,;$$

moreover k, or the time required by light to traverse the radius of the earth's orbit is $493^s,2$, and thus we have

$$k\frac{d\odot}{dt} = 20'',255.$$

We have, therefore, for the annual aberration of the fixed stars in right ascension and declination, the formulæ

$$\left.\begin{aligned}\alpha' - \alpha &= -20'',255\,[\cos\odot\cos\epsilon\cos\alpha + \sin\odot\sin\alpha]\sec\delta\\ \delta' - \delta &= +20'',255\cos\odot[\sin\alpha\sin\delta\cos\epsilon - \cos\delta\sin\epsilon]\\ &\quad -20'',255\cos\alpha\sin\delta\sin\odot\end{aligned}\right\}\ \ldots (A).$$

The terms of the second order are so insignificant that they may almost always be neglected. For the right ascension these terms will be, if we introduce into the second term of the formula (a) the values of the differential coefficients (b),

$$-\frac{1}{4}\frac{R^2}{\mu^2}\left(\frac{d\odot}{dt}\right)^2\sec^2\delta\,[\cos 2\odot\sin 2\alpha\,(1+\cos^2\epsilon) - 2\sin 2\odot\cos 2\alpha\cos\epsilon],$$

where the small term multiplied into $\sin 2\alpha\sin^2\epsilon$ is neglected†. If we substitute the numerical values, taking $\epsilon = 23^0.28'$, we obtain

$$-0'',0009155\sec^2\delta\sin 2\alpha\cos 2\odot$$
$$+0'',0009123\sec^2\delta\cos 2\alpha\sin 2\odot.$$

* In strictness, the true motion of the sun in his elliptic orbit should be used for the calculation of the aberration, but for this case the circular orbit may properly be employed instead of the elliptic, since the difference only introduces a constant term into the aberration which remains mixed up with the mean places of the stars. See Bessel's *Tabulæ Regiomontanæ*, page XIX.

† We obtain, namely, paying no regard to the factors standing before the brackets

$$2\sin 2\alpha\,[\cos^2\odot\cos^2\epsilon - \sin^2\odot] - 2\sin 2\odot\cos\epsilon\,[\cos^2\alpha - \sin^2\alpha]$$

Since now $\cos^2\alpha - \sin^2\alpha = \cos 2\alpha$, $\cos^2\odot = \frac{1}{2}(1+\cos 2\odot)$, and $\sin^2\odot = \frac{1}{2}(1-\cos 2\odot)$, we obtain the expression given above.

These terms give in the first place for stars whose declination amounts to $85\frac{1}{2}^0$, one hundredth of a second of time; they may therefore be neglected in all cases except for the pole-star.

For the declination, the terms of the second order give, if we neglect the terms which are not multiplied into tan δ*,

$$-\frac{1}{8}\cdot\frac{R^2}{\mu^2}\left(\frac{d\odot}{dt}\right)^2 \tan\delta\,\{\cos 2\odot\,[\cos 2\alpha\,(1+\cos^2\epsilon)-\sin^2\epsilon]$$
$$+2\sin 2\alpha\sin 2\odot\cos\epsilon\},$$

or,

$$+[0'',000394-0'',0004578\cos 2\alpha]\tan\delta\cos 2\odot$$
$$-0'',0004561\tan\delta\sin 2\alpha\sin 2\odot,$$

and these terms also do not amount to more than one hundredth of a second for declinations smaller than $87^0{\cdot}9$.

If, instead of the equator, we take the ecliptic for the fundamental plane, the formula (*b*) will be simpler, namely,

$$\frac{dx}{dt}=+R\sin\odot\frac{d\odot}{dt},$$

$$\frac{dy}{dt}=-R\cos\odot\frac{d\odot}{dt},$$

$$\frac{dz}{dt}=0.$$

Substituting these expressions in the formulæ (*a*), and putting λ and β in place of α and δ, we obtain for the annual aberration of the fixed stars in longitude and latitude the formulæ,

$$\left.\begin{array}{l}\lambda'-\lambda=-20'',255\cos(\lambda-\odot)\sec\beta\\ \beta'-\beta=+20\ \ ,255\sin(\lambda-\odot)\sin\beta\end{array}\right\}\ldots\ldots\ldots\ldots(B).$$

EXAMPLE. For April 1, 1849, we have for Arcturus,

$$\alpha=14^h.8^m.48^s=212^0.12'.0'',\ \ \delta=+19^0.58',1,\ \ \odot=11^0.37',2,$$
$$\epsilon=23^0.27',4.$$

* The second term of the expression for $\delta'-\delta$ in equation (*a*) multiplied into tan δ gives, namely, if we disregard the constant factor,

$$\sin^2\odot\sin^2\alpha+\cos^2\odot\cos^2\epsilon\cos^2\alpha+\tfrac{1}{2}\sin 2\odot\sin 2\alpha\cos\epsilon.$$

If now we express the squares of the sine and cosine of $\odot$ and α by $\cos 2\odot$ and $\cos 2\alpha$, and neglect the constant terms $1+\cos^2\epsilon-\cos 2\alpha\sin^2\epsilon$, we obtain the expression given above.

With these data we find,

$$\alpha' - \alpha = +18'',70,$$
$$\delta' - \delta = -\ \ 9\ ,56,$$

and since $\lambda = 202^\circ.8',\ \beta = +30^\circ.50',$

also
$$\lambda' - \lambda = +23'',19,$$
$$\beta' - \beta = -\ \ 1\ ,89.$$

14. For the purpose of simplifying the computation of the aberration in right ascension and declination, which, according to the formulæ now given, is somewhat inconvenient, tables have been constructed. The most convenient are those given by Gauss.

Gauss assumes

$$20'',255 \sin \odot = a \sin (\odot + A),$$
$$20\ ,255 \cos \odot \cos \epsilon = a \cos (\odot + A),$$

and obtains then simply:

$$\alpha' - \alpha = -a \sec \delta \cos (\odot + A - \alpha)$$
$$\delta' - \delta = -a \sin \delta \sin (\odot + A - \alpha) - 20'',255 \cos \odot \cos \delta \sin \epsilon$$
$$= -a \sin \delta \sin (\odot + A - \alpha) - 10'',128 \sin \epsilon \cos (\odot + \delta)$$
$$- 10'',128 \sin \epsilon \cos (\odot - \delta).$$

By these formulæ the tables are constructed. The first table gives, with the argument sun's longitude, A and $\log a$, from which are obtained the aberration in right ascension and the first part of the aberration in declination. The second and third parts are obtained from the second table which is entered with the arguments $\odot + \delta$ and $\odot - \delta$. These tables were first given by Gauss in the *Monatliche Correspondenz*, Vol. XVII, page 312. The constant there used is that which has been previously given, namely, 20″,255. More recently the same tables have been recomputed by Nicolai with the constant 20″,4451, and are printed in the Warnstorf Collection of Auxiliary Tables.

For the preceding example we obtain from the tables last mentioned,

$$A = 1^\circ . 1', \qquad \log a = 1{\cdot}2748,$$

and with these data

$$\alpha' - \alpha = +18'',88,$$

and the first part of the aberration in declination is $-2'',15$. In the same manner we find for the second and third parts $-3'',47$ and $-4'',03$, if the second table be entered with the arguments $31^\circ.35'$ and $-8^\circ.21'$. Thus we have

$$\delta' - \delta = -9'',65.$$

Multiplying these values of $\alpha' - \alpha$ and $\delta' - \delta$ by $\dfrac{20'',2550}{20'',4451}$, we obtain as before

$$\alpha' - \alpha = +18'',7, \text{ and } \delta' - \delta = -9'',56.$$

Besides these general tables for the aberration there are given in the *Astronomische Jahrbuch* special tables, which are arranged in the order of the days of the year. There is assumed, namely,

$$\begin{aligned} -20'',255 \cos \odot \cos \epsilon &= h \sin H, \\ -20\ ,255 \sin \odot \qquad &= h \cos H, \\ -20\ ,255 \cos \odot \sin \epsilon &= h \tan \epsilon \sin H = i, \end{aligned}$$

and we have then

$$\begin{aligned} \alpha' - \alpha &= h \sin (H + \alpha) \sec \delta, \\ \delta' - \delta &= h \cos (H + \alpha) \sin \delta + i \cos \delta. \end{aligned}$$

Such tables, which give the values of h, H, and i for every tenth or twentieth day, are found in Encke's *Jahrbuch*. For the example previously employed

$$h = +18'',65, \quad H = 257^\circ.22', \quad i = -7'',90,$$

with which we find for $\alpha' - \alpha$ and $\delta' - \delta$ the same values as before.

15. The maximum and minimum of aberration in longitude take place when the longitude of the star is equal to that of the sun or greater by 180°; on the contrary the maximum or minimum in latitude occurs when the star precedes the sun by 90° or follows it by the same quantity. The formulæ for the

Annual Parallax of the fixed stars (that is, for the angle which lines drawn from the sun and the earth subtend at the fixed stars) are strictly analogous to those for the annual aberration, with this exception, that in this case the maxima and minima occur at other times. If, for instance, Δ be the distance of a fixed star from the sun, λ and β its longitude and latitude as seen from the sun, then are the co-ordinates of the star referred to the sun:

$$x = \Delta \cos\beta \cos\lambda, \quad y = \Delta \cos\beta \sin\lambda, \quad z = \Delta \sin\beta.$$

The co-ordinates of the star referred to the earth will be

$$x' = \Delta' \cos\beta' \cos\lambda', \quad y' = \Delta' \cos\beta' \sin\lambda', \quad z' = \Delta' \sin\beta',$$

and, since the co-ordinates of the sun referred to the earth are

$$X = R\cos\odot \text{ and } Y = R\sin\odot,$$

we have

$$\Delta' \cos\beta' \cos\lambda' = \Delta \cos\beta \cos\lambda + R\cos\odot,$$
$$\Delta' \cos\beta' \sin\lambda' = \Delta \cos\beta \sin\lambda + R\sin\odot,$$
$$\Delta' \sin\beta' = \Delta \sin\beta.$$

Hence we easily obtain

$$\lambda' - \lambda = -\frac{R}{\Delta'}\sin(\lambda - \odot)\sec\beta \times 206265,$$

$$\beta' - \beta = -\frac{R}{\Delta'}\cos(\lambda - \odot)\sin\beta \times 206265;$$

or, since $\frac{R}{\Delta'} \times 206265$ is the annual parallax π,

$$\left.\begin{aligned} \lambda' - \lambda &= -\pi\sin(\lambda - \odot)\sec\beta \\ \beta' - \beta &= -\pi\cos(\lambda - \odot)\sin\beta \end{aligned}\right\} \ldots\ldots\ldots\ldots\ldots\ldots (C).$$

The formulæ are therefore altogether similar to those for the aberration, only that the maximum and minimum of parallax occur when the star precedes or follows the star by 90^0; on the other hand, the maximum or minimum in latitude occurs when the longitude of the star is equal to that of the sun, or is greater by 180°.

For the right ascension and declination we have the equations

$$\Delta' \cos\delta' \cos\alpha' = \Delta \cos\delta \cos\alpha + R \cos\odot,$$
$$\Delta' \cos\delta' \sin\alpha' = \Delta \cos\delta \sin\alpha + R \sin\odot \cos\epsilon,$$
$$\Delta' \sin\delta' = \Delta \sin\delta + R \sin\odot \sin\epsilon,$$

from whence, in the same manner as for the aberration, we find

$$\left.\begin{aligned} \alpha' - \alpha &= -\pi\,[\cos\odot \sin\alpha - \sin\odot \cos\epsilon \cos\alpha] \sec\delta \\ \delta' - \delta &= -\pi\,[\cos\epsilon \sin\alpha \sin\delta - \sin\epsilon \cos\delta] \sin\odot \\ &\quad - \pi \cos\odot \sin\delta \cos\alpha \end{aligned}\right\} \ldots (D).$$

16. The daily motion of the earth on its axis produces in the same manner as the yearly motion round the sun an *Aberration*, which is called the *Diurnal Aberration*. This is however much more insignificant than the annual aberration, since the velocity of the motion of the earth on its axis is very much smaller than the velocity of the motion in the annual orbit round the sun.

The co-ordinates of a place on the surface of the earth referred to three rectangular co-ordinates of which one coincides with the axis of rotation, and the two others lie in the plane of the equator, so that the positive axis of x is drawn from the center towards the vernal equinox, the positive axis of y towards the ninetieth degree of right ascension, are, according to No. 2 of this section,

$$x = \rho \cos\phi' \cos\theta,$$
$$y = \rho \cos\phi' \sin\theta,$$
$$z = \rho \sin\phi'.$$

We have thus

$$\frac{dx}{dt} = -\rho \cos\phi' \sin\theta \frac{d\theta}{dt},$$
$$\frac{dy}{dt} = +\rho \cos\phi' \cos\theta \frac{d\theta}{dt},$$
$$\frac{dz}{dt} = 0.$$

Substituting these values in the formulæ (a) in No. 13, we easily obtain, neglecting the higher powers,

$$\alpha' - \alpha = \frac{1}{\mu}\frac{d\theta}{dt} . \rho \cos\phi' \cos(\theta - \alpha) \sec\delta,$$

$$\delta' - \delta = \frac{1}{\mu}\frac{d\theta}{dt} . \rho \cos\phi' \sin(\theta - \alpha) \sin\delta.$$

Denoting now by T the number of sidereal days contained in the time in which the sun goes through 360° in the heavens, or the so-called sidereal year*, then is the angular motion of a point of the earth's surface on account of its rotation T times greater than the angular motion of the earth in its orbit, so that

$$\frac{d\theta}{dt} = T\frac{d\odot}{dt}.$$

Hence therefore we obtain as the constant of diurnal aberration, (since

$$\frac{1}{\mu}\rho = k\frac{\rho}{R} = k \sin\pi,$$

where π is the sun's parallax and k the number of seconds of time required by light to traverse the radius of the earth's orbit,)

$$k\frac{d\odot}{dt}\sin\pi . T;$$

or, (since $k\frac{d\odot}{dt} = 20'',255$, $\pi = 8'',5712$, and $T = 366{\cdot}26$) it is equal to 0'',3083.

If, besides, instead of the corrected latitude ϕ' the latitude ϕ be simply substituted, we obtain for the diurnal aberration in right ascension and declination,

$$\left.\begin{aligned}\alpha' - \alpha &= 0'',3083 \cos\phi \cos(\theta - \alpha) \sec\delta \\ \delta' - \delta &= 0\ \ ,3083 \cos\phi \sin(\theta - \alpha) \sin\delta\end{aligned}\right\} \cdots\cdots\cdots (E).$$

Whence it follows that, on the meridian, the diurnal aberration

* This time is, as will be seen in the sequel, somewhat greater than the time which elapses between two passages of the sun through the vernal equinox, since the sidereal year = 365·25637 mean solar days; or is equal to 365 days, 6 hours, 9 minutes, and 10·7496 seconds.

of the stars in declination is nothing, while, in right ascension, it attains its maximum, namely,

$$0'',3083 \cos \phi \sec \delta.$$

17. For the annual aberration of the fixed stars in longitude and latitude the following expressions have been already found:

$$\lambda' - \lambda = - k \cos (\lambda - \odot) \sec \beta,$$
$$\beta' - \beta = + k \sin (\lambda - \odot) \sin \beta,$$

where the constant 20″,255 is denoted by k. Imagine now at the mean place of the star a tangent plane to the apparent sphere of the heavens, and a rectangular pair of co-ordinates at this point, whose axes of x and y are the lines of intersection of the circles of parallel and of latitude with the tangent plane, and refer now the true place affected with aberration to the mean place by the co-ordinates

$$x = (\lambda' - \lambda) \cos \beta, \text{ and } y = \beta' - \beta^*,$$

then we easily obtain, by squaring the above equations,

$$y^2 = k^2 \sin^2 \beta - x^2 \sin^2 \beta.$$

Now this is the equation to an ellipse, whose semi-major axis is equal to k and the semi-minor axis to $k \sin \beta$. Thus, on account of annual aberration, the fixed stars describe round their mean place an ellipse whose semi-major axis is 20″,255, and the semi-minor axis is the maximum of the aberration in latitude. For stars which are in the ecliptic $\beta = 0$, and consequently the semi-minor axis is equal to 0. Such stars therefore describe in the course of a year a straight line, receding along the ecliptic on each side of the mean place, by the quantity 20″,255. For a star which is in the pole of the ecliptic, $\beta = 90^\circ$, and consequently the semi-minor axis is equal to the semi-major axis. Such a star will thus in the course of a year describe about its mean place a circle whose radius is 20″,255.

The same reasoning precisely applies to the annual parallax and the diurnal aberration. By means of the latter the stars, in

* Since for such minute distances from the origin of co-ordinates the tangent plane may be considered to coincide with this spherical surface.

the course of a sidereal day, describe about their mean place ellipses whose semi-major and semi-minor axes are respectively

$$0'',3083 \cos \phi, \text{ and } 0'',3083 \cos \phi \sin \delta.$$

For stars in the equator this ellipse becomes a straight line, and for a star at the pole it becomes a circle.

18. If the body have a proper motion, like the sun, the moon, and the planets, then for such the aberration of the fixed stars before treated of, is not the complete aberration. For since such a body changes its place during the time taken by the ray of light to traverse the space between it and the earth, so the observed direction of the ray does not correspond to the true geocentric place of the body at the time of observation. Let us suppose that the ray of light which reaches the object-glass of the telescope at the time t, set out from the body at the time T. Let also P (fig. 4) be the place of the planet in space at the time T, p the same at the time t, A the position of the object-glass at the time T, a and b the positions of the object-glass and the eye-piece at t, and a' and b' their positions at the time t', when the ray reaches the eye-piece. Then we have:

1. AP the direction towards the place of the planet at the time t.
2. ap the direction towards the true place at the time t.
3. ap or $a'p'$ the direction towards the apparent place at the time t, or at the time t', whose difference is indefinitely small.
4. $b'a$ the direction towards the same apparent place, freed from the aberration of the fixed stars.

Since now P, a, and b' lie in a straight line, we have

$$Pa : ab' = t - T : t' - t.$$

Since moreover the interval of time $t' - T$ is always very small, so that we may assume that within its limits the earth moves in a straight line and with a uniform velocity, the points A, a, and a' also lie in a straight line, so that Aa and aa' are also proportional to the times $t - T$ and $t' - t$. From hence it follows that AP is parallel to $b'a'$, and that thus the apparent place of the planet at the time t is the same as the true place at the time T. But the difference of the times t' and T is the time in which the light from the planet reaches the eye, or the product of the distance of the planet into $493^s,2$, that is, into the time in which light traverses the semi-major axis of the earth's orbit, which is taken as the unit.

Hence there result three methods for computing the true place of a planet from the apparent at any time t.

1. Subtract from the observed time the time in which the light from the planet reaches the earth; we then obtain the time T, and the true place at the time T is identical with the apparent place at the time t.

2. Compute with the distance of the planet the reduction of time $t - T$, and with this by help of the daily motion of the body in right ascension and declination the reduction of the observed apparent place to the time T.

3. Consider the given place freed from the aberration of the fixed stars as the true place at the time T, but as seen from the place which the earth occupies at the time t.

This last method is to be employed when the distance of the body is not known, for example, in the computation of an orbit of a yet unknown planet or comet.

Since the time in which the light from the sun reaches the earth is $493^s,2$, and the mean motion of the sun in a day is $59'.8'',3$, we have (from 2) the aberration of the sun in longitude equal to $20'',25$, by which quantity the longitude of the sun is always observed too small. On account of the change of distance and velocity of the sun this value varies in the course of a year by some tenths of a second.

NOTE. On the subject of Aberration compare the Preface to Bessel's *Tabulæ Regiomontanæ*, page xvii. &c., and Gauss, *Theoria Motûs*, page 68, &c.

THIRD SECTION.

DETERMINATION OF THE CO-ORDINATES AND ANGLE OF THE APPARENT SPHERE OF THE HEAVENS INDEPENDENT OF THE POSITION OF THE OBSERVER ON THE SURFACE OF THE EARTH. PERIODICAL AND SECULAR CHANGES OF THESE QUANTITIES.

THE co-ordinates and angle of the apparent sphere of the heavens which are independent of the position of the observer on the surface of the earth are the right ascensions and declinations and the longitudes and latitudes of the stars, and finally the angle which the fundamental planes of the two systems of co-ordinates make with one another, or the obliquity of the ecliptic. The spherical co-ordinates of longitude and latitude are never determined immediately by observations, but are always deduced by computation, by means of the formulæ for the transformation of co-ordinates, from right ascensions and declinations (I. No. 8). There remain therefore to be determined by observation only the right ascensions and declinations and the obliquity of the ecliptic.

By comparing with each other at different epochs the determinations of these quantities, it is found that they are subject to changes of which one part, in intervals of time not very great, is proportional to the time, while the other is periodical. The change proportional to the time of the right ascension and declination as well as of the longitude and latitude is called the *Precession;* and on the other hand the change proportional to the time of the obliquity of the ecliptic is called the *Secular Variation of the Obliquity.* The other part of the change whose principal terms have a period of 19 years, is denoted by the term *Nutation*. Both changes have their origin in a secular motion of the equator on the ecliptic as well as of the ecliptic on the equator, whereby the inclination of the two planes with respect to each other is altered; and in a periodical oscillation of the intersection of the equator and ecliptic on the latter plane, as well as in a periodical change of the inclination of the ecliptic and equator connected with the same.

The place of a star at any given time when freed from the periodical part of the change or the Nutation, is called the *mean place of the star* for that epoch. These mean places of the stars are given in star-catalogues. To obtain from thence the mean places for any other epoch, application must be made of the precession for the difference of the times; but, if it be required to find the true place of the star referred to the true equinox for this time, it is necessary to add the nutation as well as the precession. It is therefore necessary to find means for knowing the law of the changes of the places of the stars on account of precession and nutation, and at the same time to devise convenient methods both for reducing the mean places of the stars to different epochs, as also for changing the mean places into true places and *vice versâ*.

I. Determination of the Right Ascensions and Declinations of the Stars and of the Obliquity of the Ecliptic.

1. If we observe the difference of the times at which stars pass the meridian of a place, these differences are also the differences of right ascension of the stars expressed in time. (I. No. 3. Note.) If also at the same time there be observed the altitudes of the stars at their meridian passages, we obtain also the differences of their declinations, since every meridian altitude of a star differs from its declination by a constant quantity. (I. No. 14.)

For these observations there is necessary a good clock (that is, such a one as for equal arcs of the equator passing across the meridian give an equal number of seconds*) and an altitude

* It is not necessary to know the absolute time, since only differences of time are observed.

instrument fixed in the plane of the meridian, that is, a meridian circle. This in its essential parts consists of a horizontal axis lying in two fixed Y's, which carries a vertical circle and a telescope. To one of the supports is fastened an index, which, by means of the simultaneous motion of the telescope and the circle round the horizontal axis, gives on the circle the arc passed over by the telescope.

To test the uniform rate of going of the clock, consecutive transits are observed of different stars on a vertical wire stretched in the focus of the telescope. If then the instrument have not changed its position in the interval, and if the observation be made on the same part of the vertical wire, then must the clock, if it be adjusted to sidereal time, shew 24 hours exactly between two consecutive transits of the same star. If this be not the case, but the clock gives, for consecutive transits of any star, the time $24^h - \alpha$, then is α called the daily *rate* of the clock and must in the observation of differences of right ascension be taken into account, by multiplying the observed difference by

$$\frac{24}{24 - \alpha}.$$

When the uniformity of rate of the clock has been ascertained, it is necessary so to adjust the meridian circle that the vertical wire of the telescope may be in the plane of the meridian for every position of the telescope. If the axis of the instrument has been made horizontal by means of a spirit-level, then a star near the equator is allowed when near the meridian* to run along another wire placed at right angles to the first, and the plate carrying the wires is turned till the star as long as it is in the field runs well along the wire. Then is this wire accurately horizontal and the other is accurately vertical. After this has been accomplished the telescope is directed to a distant terrestrial object, and a distinguishable point is noted which is bisected by the vertical wire. Then the instrument must be shifted in its Y's so that the circle which was in the east side before may be now on the west side, and the telescope must be directed in this position to the same object. If then the vertical wire in this

* The direction of the meridian, as far as this object is concerned, can be found with sufficient accuracy by observing the time at which the altitudes of the stars do not change.

position also bisect with sufficient accuracy the same point of the object, the line of sight, that is, the line from the center of the object-glass to the wire-cross is sensibly perpendicular to the axis of rotation of the instrument. But if the wire bisect another point, then the wire frame must be shifted by means of screws which give it a motion at right angles to the line of sight, till the vertical wire passes accurately through the point of the object that lies midway between the two points bisected in the reversed positions. The line of sight will be now at right angles to the axis of rotation. If this be still not accurately the case, the operation can be repeated till the error be totally got rid of.

Finally, to bring the vertical wire into the plane of the meridian, use must be made of the pole star, by observing the transits over the wires at three consecutive and opposite culminations. If for instance the instrument be accurately in the plane of the meridian, the time between the upper culmination and the lower culmination next following must be accurately equal to the time which elapses between the lower and the next following upper culmination. If this be not the case, we know that the vertical circle which the instrument describes deviates from the meridian on that side of it in which the star is for the shorter period of time, and the line of sight can be brought into the plane of the meridian by motion of one of the Y's of the instrument. In this manner can the adjustments of such an instrument be accurately performed, so that the vertical wire of the telescope in each position may be accurately in the plane of the meridian.

After this has been accomplished, the times of transit of stars over the vertical wires must be observed, and at the same time, a little before or after the meridian passage, they must be brought upon the horizontal wire, and the number must be read off which is shewn by the index on the circle in this position of the telescope. We then obtain from the differences of the observed times the differences of the right ascensions, and, from the differences of the readings of the circle, the differences of apparent declinations. To these observed differences are still to be applied the corrections treated of in the preceding sections, for the purpose of deducing the true differences of right ascension and declination of the stars.

The parallax in right ascension is nothing on the meridian, and therefore is not to be regarded in the observation of the differences of right ascension; on the contrary the declination, or rather the reading of the circle, must be freed from the parallax, when the body observed is affected by it. If the divisions increase in the direction of the zenith-distance, and thus increase from the zenith towards the horizon, $-\pi \sin z$ must be applied to the circle-readings, (II. No. 3) where π is the horizontal parallax and z the apparent zenith-distance of the object*. For the fixed stars this correction is nothing. Since moreover the refraction always acts only in the vertical direction, it does not alter the times of transit of the stars over the meridian; but to all circle results, on the contrary, it is necessary, if the divisions increase in the direction of zenith-distance, to apply the correction $+r$, where r is to be computed according to formula (F) in No. 10 of the second section, and account is to be taken of the readings of the barometer and thermometer at the time of observation.

Since both corrections require a knowledge of the apparent zenith-distance, it is necessary for the purpose of being able to compute them from the circle-readings, to know what point of the same corresponds to the zenith. This point, which is called the *zenith point* of the circle, is easily found by placing the horizontal wire of the telescope in two different positions of the instrument (that is with the circle East and West) on the same terrestrial object. If ζ be the reading of the circle in one observation and ζ' in the other, then is the zenith point $\frac{1}{2}(\zeta + \zeta')$. Instead of a terrestrial object, use may be made of the pole-star at the time when it is very near the meridian, since at this time the zenith-distance changes with extreme slowness.

In the last place the observed differences of right ascension and declination are to be freed from aberration, by applying to the observations the expressions (A) given in II, No. 13; for right ascensions, to the observed times with signs changed, and on the contrary the correction $\delta' - \delta$ with proper sign to the

* If the divisions increase contrary to the direction of zenith-distance from the horizon to the zenith, all the corrections are to be applied with contrary signs.

observed circle-readings when the stars pass the meridian south of the zenith, and the divisions increase in the direction of the zenith-distances. Since these expressions for $\alpha' - \alpha$ and $\delta' - \delta$ themselves contain the quantities α, δ, and ϵ, their computation presupposes an approximate knowledge of them. This we have by means of earlier star-catalogues. In former times the ancients determined the right ascensions and declinations of the stars naturally without reference to these small corrections, but by means of a method which was essentially the same as that which is employed at present. Since that time the catalogues have been receiving continual corrections, in part by the superior accuracy of the observations since the invention of the telescope and of the wire-micrometer, and partly by the employment of more accurate values of the small corrections.

If the object have a visible disk, as for example the sun, then must its semi-diameter be applied to the observation of zenith-distance, or otherwise the upper as well as the lower limb must be observed on the meridian. In these cases the refraction must be applied to the observation of each limb separately, and the mean of the corrected zenith-distances must then be taken.

After the differences of true right ascension and declination of the stars have been thus found, the only thing still necessary is to determine the true right ascension and declination of a single star, or rather the true right ascension of a star and that point of the meridian circle which corresponds to the pole or to the height of the equator, in order to obtain the right ascensions and declinations of all the other stars. If now these determinations be made at different times, it is found that, leaving out of account errors of observations, the right ascensions and declinations are never the same on account of the changes in the positions of the planes to which the star's places are referred, which cause the stars apparently to change their places with reference to the planes. These changes however cannot be taken into consideration before the place proper for their introduction.

2. The point of the circle, which corresponds to the pole, technically called the polar point, is found by observations of

circumpolar stars at their upper and lower culminations. If for instance ζ be the reading of the circle corrected for refraction for the upper culmination, and ζ' for the lower culmination, then is $\frac{1}{2}(\zeta' - \zeta) = 90^0 - \delta$, and $\frac{1}{2}(\zeta' + \zeta)$ is the point of the circle corresponding to the pole. On the other hand, $\frac{1}{2}(\zeta' + \zeta) \pm 90^0$ according to the direction of the divisions of the circle is the point which corresponds to the altitude of the equator, or, to the equatorial point of the circle. If also, the zenith-point Z of the circle be known by the methods before pointed out, then is

$$Z - \frac{1}{2}(\zeta' + \zeta), \text{ or } \frac{1}{2}(\zeta' + \zeta) - Z$$

the co-latitude of the place of observation. After the polar point of the circle has been determined, the declinations of all observed objects can be found, and it only remains to deduce from observation, a true right ascension of a single object.

Since now there is taken for the zero of right ascensions of the stars the point in which the ecliptic (that is, that great circle which the sun in the course of a year appears to describe on the visible sphere of the heavens) cuts the equator, we shall obtain a knowledge of the right ascension of a star by connecting the observations of the culmination of the stars with those of the sun. If for instance for several successive days about the times of the equinoxes, besides the culminations of the sun and of a star, there be also observed the declinations of the center of the sun, we obtain for different declinations of the sun the differences of right ascension of the sun and the star, and can therefore compute this difference for the instant when the declination of the sun is equal to 0^0, and therefore the right ascension is either 0^0 or 180^0. If then the observations be made at the vernal equinox, the computed difference of right ascension will be the absolute right ascension of the star, and on the other hand, if the observations be made at the autumnal equinox, we shall find a value differing from the right ascension by 180^0.

The third of the quantities to be determined is the obliquity of the ecliptic, or the angle which the plane of the ecliptic makes with that of the equator. The measure of this angle is the arc of the solstitial colure (that is, of the circle of latitude passing through the poles of both great circles) which is comprised between the equator and the ecliptic. The obliquity of the ecliptic is therefore also equal to the greatest declination which the center of the sun has in the course of the year. Therefore if we observe about the time of the summer solstice (June 21) for every day the declination which the sun has at its transit over the meridian, then, if the time of a culmination be coincident with the time of the solstice, the greatest of the observed declinations is immediately the obliquity of the ecliptic. But if this be not the case, the greatest declination can be easily deduced from those observed, by finding the time for which the first difference of the observed declinations is equal to nothing, and by interpolating the declination for this time.

At the expiration of half a year, at the time of the winter solstice, if the sun were again observed, the same absolute value would be found for the greatest southern declination of the sun if the observations were without error*. In this case, moreover, when both solstices are observed, the knowledge of the polar point of the circle is not at all requisite, but only the zenith-point, nor is (which amounts to the same thing) a knowledge of the latitude of the place of observation necessary.

If, for instance, for the least zenith-distance of the sun's center in summer the value z were found, and for the greatest zenith-distance in winter the value z', then is $\frac{1}{2}(z'-z)$ equal to the obliquity of the ecliptic and $\frac{1}{2}(z'+z)$ is equal to the zenith-distance of the equator or the latitude.

Every two observations of the difference of right ascension of the sun and a star and of the declination of the sun give moreover both the right ascension of the star and the obliquity of the ecliptic. For instance, if α be the right ascension of the star, A the observed difference of right ascension of the sun and

* With the exception of a trifling difference arising from the secular change of the obliquity and the nutation.

star*, D the declination of the sun, and ϵ the obliquity of the ecliptic, we have, according to the Note to No. 9 of Section I,

$$\sin (A + \alpha) \tan \epsilon = \tan D,$$

and, in like manner, from the second observation,

$$\sin (A' + \alpha) \tan \epsilon = \tan D'.$$

From the two observations we find

$$\sin \alpha \tan \epsilon = \frac{\tan D \sin A' - \tan D' \sin A}{\sin (A' - A)},$$

and

$$\cos \alpha \tan \epsilon = \frac{\tan D' \cos A - \tan D \cos A'}{\sin (A' - A)},$$

from which α and ϵ can both be computed.

At the same time it is always desirable to determine the values of the quantities α and ϵ independently of each other, so that error of the one may not seriously affect the other, and we must for this purpose proceed by a method similar to those previously given.

3. Assuming that the position of the vernal equinox is approximately known by one of the previous methods, the obliquity of the ecliptic can be rigorously determined from observations of the sun in the neighbourhood of the solstices, in the following manner. If x be the distance of the sun in right ascension from the solstice, which is thus equal to $90^\circ - \alpha$, we have the equation

$$\cos x \tan \epsilon = \tan D.$$

Since, by the supposition, x is a small quantity, ϵ can from this equation be developed in a rapidly converging series, since by formula (20) of the Introduction we obtain

$$\epsilon = D + \tan^2 \frac{x}{2} \sin 2D + \frac{1}{2} \tan^4 \frac{x}{2} \sin 4D + \&c. \ldots\ldots\ldots (A).$$

In this manner can the obliquity of the ecliptic be obtained from one observation of the declination of the sun in the neighbourhood of the solstices.

* So that $A + \alpha$ is the right ascension of the sun.

Bessel observed at Konigsberg, when the right ascension of the sun was $5^h.51^m.23^s,5$,

$$D = 23^\circ.26'.47'',83.$$

Since the right ascension of the sun at the time of the solstices is 6 hours, we have here

$$x = 8^m.36^s,5 = 2^\circ.9'.7'',5.$$

We have thus

$$\tan^2\frac{x}{2}\sin 2D = +53'',13,$$

$$\frac{1}{2}\tan^4\frac{x}{2}\sin 4D = +\ \ 0,01,$$

and therefore from this observation we have the obliquity of the ecliptic,

$$\epsilon = 23^\circ.27'.40'',97.$$

Now for the purpose of freeing the result from casual errors of observation, the declination should be observed on several days in the neighbourhood of the solstice, and the means of the separate values of ϵ thus obtained, should be taken. The time of the solstice is necessary to be known only approximately, since an error in x produces only a very small effect on the determination of ϵ. For example, if we take into account only the first term of the series,

$$d\epsilon = \frac{\tan\frac{x}{2}\sin 2D}{\cos^2\frac{x}{2}}\,dx,$$

or, from the fundamental equation,

$$d\epsilon = \frac{1}{2}\tan x \sin 2\epsilon\, dx;$$

so that, for example, we should have only an error of $1'',37$ in ϵ, if the assumed value of x were in error to the amount of $100''$.

4. If then the obliquity of the ecliptic be known, the absolute right ascension of a star can be determined with the utmost accuracy. A bright star is selected, which can be observed in

the daylight as well as by night, and which is in the neighbourhood of the equator. Ordinarily, Altair (α Aquilæ) or Procyon (α Canis Minoris) is selected. Then, in the first place, every observation of the sun gives (if A now denote the true right ascension of that body) the equation

$$\sin A \tan \epsilon = \tan D,$$

or

$$A = \sin^{-1} \frac{\tan D}{\tan \epsilon}.$$

Now let the star be observed on the meridian at the clock time t, the sun at the clock time T, then is the right ascension α of the star equal to

$$\alpha = \sin^{-1} \frac{\tan D}{\tan \epsilon} + (t - T)^{*}.$$

By this equation the right ascension of the star is thus found from the observed difference of right ascension of the star and the sun, the declination of the latter being D, and the obliquity of the ecliptic being ϵ. If therefore D and ϵ be in error†, we shall on this account also obtain an erroneous value of α, independently of errors of observation in $t - T$. But, by differentiating logarithmically the equation

$$\sin A \tan \epsilon = \tan D,$$

we obtain

$$\cot A dA + \frac{2d\epsilon}{\sin 2\epsilon} = \frac{2dD}{\sin 2D},$$

and consequently, if we add these terms to the equation for α,

$$\alpha = t - T + \sin^{-1} \frac{\tan D}{\tan \epsilon} + \frac{2 \tan A}{\sin 2D} dD - \frac{2 \tan A}{\sin 2\epsilon} d\epsilon.$$

Now for the purpose of obtaining α independently of the errors dD and $d\epsilon$, several observations must be combined in such a way, that these errors shall eliminate each other's effects. This will be effected by combining an observation near the

* Neglecting the rate of the clock in the interval $t - T$. TRANSLATOR.

† Properly speaking it is only a constant error in D that is to be taken into consideration here, since casual errors of observations will be got rid of in a mass of observations.

vernal equinox with another near the autumnal equinox. For instance, if in the equation

$$\sin A = \frac{\tan D}{\tan \epsilon}$$

we take the angle A always acute, we have for the latter observation the equation

$$\alpha = t' - T' + \left(180^\circ - \sin^{-1}\frac{\tan D'}{\tan \epsilon}\right) - \frac{2\tan A'}{\sin 2D'}\,dD + \frac{2\tan A'}{\sin 2\epsilon}\,d\epsilon,$$

and from the mean of the two equations we obtain for α

$$\alpha = \frac{1}{2}\left[\overline{t - T} + \overline{t' - T'}\right] + \frac{1}{2}\left(\sin^{-1}\frac{\tan D}{\tan \epsilon} - \sin^{-1}\frac{\tan D'}{\tan \epsilon} + 180^\circ\right)$$

$$+ \left(\frac{\tan A}{\sin 2D} - \frac{\tan A'}{\sin 2D'}\right) dD - \frac{\tan A - \tan A'}{\sin 2\epsilon}\,d\epsilon \;\ldots.\; (B).$$

If now the acute angle $A' = A$, then D' is also equal to D; and if therefore the difference of right ascensions of the sun and the star be observed at the times when the sun has the right ascensions A and $180^\circ - A$, then will the coefficients of dD and $d\epsilon$ in equation (B) be nothing, and the constant errors in the declination and the obliquity will thus have no effect on the right ascension of the star. This it is true can never be attained with the utmost rigour, since it will never exactly happen that, when the sun at one culmination has the right ascension A, the right ascension $180^\circ - A$ shall exactly correspond to another culmination. But if A in the first case be only nearly equal to $180^\circ - A$ in the second, the remaining errors dependent on dD and $d\epsilon$ will be always exceedingly small.

Thus, for the determination of the absolute right ascension of a star, the difference of right ascensions of the sun and star should be observed as near as possible to the vernal and autumnal equinoxes; but if one observation be made after the vernal equinox, the second must be made before the autumnal equinox, and *vice versa*, that the sun's declination may in each case have the same sign.

Bessel observed in 1828, March 23, the declination of the sun's center, cleared from refraction and parallax in altitude,

$$D = +1^\circ.\,6'.\,54'',2,$$

and the transit across the meridian

$$T = 0^h.\,11^m.\,12^s{,}57.$$

And, on the same day, the transit of α Canis Minoris

$$t = 7^h.\,31^m.\,14^s{,}62^*.$$

In the same manner he observed on the 20th of September of the same year

$$D' = +\,1^0.\,1'.\,56'',8,$$

$$T' = 11^h.\,50^m.\,33^s{,}40,$$

$$t' = 7^h.\,30^m.\,24^s{,}82.$$

To these observed quantities the aberration must now be applied. Now for the star we obtain, according to the formula (A) in No. 13 of the second section, taking $\alpha = 112^0.\,34',3$, $\delta = +\,5^0.\,39',5$, the aberration in right ascension,

Mar. 23, $+\,0^s{,}42$,

Sep. 20, $-\,0^s{,}54$,

where the signs are so to be interpreted, that these corrections are to be applied with changed signs to apparent places, to obtain the mean places. For the sun the aberration is to be computed according to the prescript in No. 18 of the second section. Now the hourly motion of the sun in right ascension is

Mar. 23, $+\,9^s{,}08$,

Sep. 30, $+\,9\,{,}00$;

and in declination

Mar. 23, $+\,59'',08$,

Sep. 30, $-\,58'',38$.

Consequently the aberration of the sun in right ascension and declination is

Mar. 23,	$+\,1^s{,}24$,	$+\,8'',09$,
Sep. 30,	$+\,1^s{,}23$,	$-\,8'',00$.

* These times are corrected for the rate of the clock.

And these corrections are to be added algebraically to the apparent places, to obtain the mean places*.

Taking account of these corrections we find

$$t - T = +7^h.20^m.\ 0^s,39$$
$$t' - T' = -4.20.\ 9,27$$

$$\frac{1}{2}[\overline{t-T} + \overline{t'-T'}] = 1.29.55,56.$$

And, if we assume ϵ to be equal to $23^{\circ}.27'.33'',4$,

$$\tan D = 8{\cdot}2901033$$
$$\tan \epsilon = 9{\cdot}6374572$$

$$\sin^{-1}\frac{\tan D}{\tan \epsilon} = 2^{\circ}.34'.32'',94$$
$$= 0^h.10^m.18^s,20,$$

$$\tan D' = 8{\cdot}2548551$$
$$\tan \epsilon = 9{\cdot}6374572$$

$$\sin^{-1}\frac{\tan D'}{\tan \epsilon} = 2^{\circ}.22'.29'',63$$
$$= 0^h.9^m.29^s,98.$$

Thus

$$\frac{1}{2}\left(\sin^{-1}\frac{\tan D}{\tan \epsilon} - \sin^{-1}\frac{\tan D'}{\tan \epsilon} + 12^h\right) = 6^h.0^m.24^s,11,$$

and finally

$$\alpha = 7^h.30^m.19^s,67.$$

In this computation the equinox has been regarded as fixed; but since this is variable on account of precession and nutation, there is still a correction to be applied to the value above found for the right ascension. The calculation of the example, taking account of this latter correction, is found in No. 11 of this section.

If the coefficients of dD and $d\epsilon$ be computed we obtain

$$\alpha = 7^h.20^m.19^s,67 + 0^s,000223\,dD + 0^s,004406\,d\epsilon.$$

* In general it is not necessary to take into account the aberration of the sun, since this only alters the time of passage through the equinox, and is eliminated by the combination of the two equations.

Thus the constant error of the declination and of the assumed obliquity of the ecliptic are very nearly eliminated by the combination of the two observations.

II. *Variations of the Planes, to which the places of the Stars are referred.*

PRECESSION AND NUTATION.

5. If by the methods above described a series of determinations were made of the points of intersection of the ecliptic and equator, it would be found that, with very few exceptions, the right ascensions of the stars increase, and, for intervals of time not very large, leaving out of account small inequalities, that their increase is proportional to the time. For different stars also a different annual change will be observed, without however any conspicuous law being observable. In the same manner if the declinations of stars be observed at different times, there will also be found in this co-ordinate a similar change proportional to the time, whose direction is different according to the quadrant in which the right ascension of the star lies. In all these changes there will be immediately discovered a remarkable law, if the stars be no longer referred to the fundamental plane of the equator but to the ecliptic. In this case namely it will be found that the longitudes of all stars increase by nearly equal quantities while their latitudes remain very nearly unchanged.

This uniform change of the places of the stars with regard to the ecliptic was first discovered by Hipparchus (B.C. 130), who compared his own observations of the stars' places with those of Timocharis, which were made about 160 years earlier. He found from this comparison that the longitudes of all the stars were changed yearly by about $36''$, and therefore in a hundred years by about 1°. This value is however too small. Hipparchus found the longitude of Spica Virginis $174^\circ.0'$; at the present time it is $201^\circ.41'$. If we take for the interval of time 1980 years, and assume the motion to be proportional to the time, we obtain for the annual motion of the equinox in longitude $50'',3$.

This change of the stars' places has its origin first in the circumstance that the point of intersection of the equator with the ecliptic has a retrograde motion on the latter plane, and, secondly, in the change of inclination of the two planes with respect to each other. The first part of this change is called the *precession* of the stars, or the *retrograde motion of the equinox;* the second is called the *secular change of the obliquity of the ecliptic.* The explanation of these phenomena belongs to Physical Astronomy, which teaches that they originate firstly in the attraction of the sun and moon on the spheroidal earth, and secondly on the action of the planets on the position of the plane of the earth's orbit. The attraction of the sun and moon does not change the inclination of the equator to the ecliptic*, but it simply causes the point of intersection of the equator and ecliptic to retrograde on the latter plane. This motion of the equator on the ecliptic is called the *Lunisolar Precession.* By it the longitudes of all stars are changed while their latitudes remain unchanged. If we take as the fixed plane that great circle of the heavens, with which the ecliptic coincided at the beginning of the year 1750, then, according to Bessel, the lunisolar precession for the year $1750 + t$ is

$$\frac{dl_1}{dt} = +50''\cdot 37572 - 0''\cdot 000243589\, t,$$

or the actual change in the interval from 1750 to $1750 + t$ is

$$l_1 = +50''\cdot 37572\, t - 0''\cdot 000121795\, t^2,$$

by which quantity the longitudes of all stars are increased in this interval.

The mutual attractions of the planets produce in addition a change of inclination of their orbits to each other, and a forward motion of their nodes; that is, of the intersections of the planes of their orbits. Since now the earth's equator is not changed by these attractions, they produce a change of the obliquity of the ecliptic, and a motion of the point of intersection of the ecliptic and equator on the latter plane. This motion of the equinoctial points is called the *Planetary Precession.* By its means the right ascensions of all stars are changed, whilst the declinations re-

* At least the changes of the inclination thereby produced are only periodical, which in the sequel will be treated of under the head of Nutation.

main the same, and, according to Bessel, the annual diminution of the right ascension for the time $1750+t$ is

$$\frac{da}{dt}=+0''\cdot17926-0''\cdot0005320788\,t^{*}.$$

If we denote by a the quantity by which the right ascensions of all stars are diminished in the interval from 1750 to $1750+t$, we have

$$a=0''\cdot17926\,t-0''\cdot0002660394\,t^2.$$

At the same time the obliquity of the ecliptic is also changed, and we have for its annual variation by means of the planets for the time $1750+t$,

$$\frac{d\epsilon}{dt}=-0''\cdot48368-0''\cdot0000054459\,t;$$

and, for the obliquity itself for the time $1750+t$, we have

$$23^{\circ}.28'.18'',0-0''\cdot48368t-0''\cdot0000027230t^2.$$

But in addition, the position of the ecliptic with respect to the equator being changed, the attraction exerted by the sun and the moon on the spheroidal earth is also changed, and a long inequality in the inclination of the plane of the equator to the ecliptic is introduced. On this account also there arises a change of the obliquity of the fixed ecliptic for 1750† with respect to the equator, whose annual variation is

$$\frac{d\epsilon_0}{dt}=+0''\cdot00001968466\,t,$$

and the obliquity of the fixed ecliptic for the time $1750+t$ is

$$\epsilon=23^{\circ}.28'.18'',0+0''\cdot00000984233\,t^2.$$

Let now (fig. 5) AA_0 be the equator and EE_0 the ecliptic, both for the year 1750; let also $A'A''$ and EE' denote the positions of the equator and ecliptic for the year $1750+t$; then is BD, the portion of the fixed ecliptic through which the equator has retrograded on the latter, or, the lunisolar precession in t

* Hence in the year 2087 the motion of the ecliptic on the equator, which is now in the contrary direction to that of the equator on the ecliptic, will be in the same direction.

† Namely, the motion of the equator with reference to the ecliptic with sign changed.

years $= l_1$; in addition the portion BC, through which the ecliptic has moved forwards on the equator, or the precession through the planets in t years, is equal to a; and lastly are BCE and $A'BE$ respectively, the inclinations of the true and the fixed ecliptic to the equator, equal to ϵ and ϵ_0. If then S be any star, and SL and SL' be drawn perpendicular to the fixed and the true ecliptic, DL is the longitude of the star for 1750, and CL' is the longitude of the star for $1750 + t$. Denote now by D' that point of the moveable ecliptic which on the fixed ecliptic is denoted by D, then is the portion CD', that is, that portion of the true ecliptic between the equinox for 1750 and the equator for the time $1750 + t$, called the *General Precession* in the time t, since this part of the precession in longitude is the same for all stars. To obtain from thence the complete precession for a star in longitude, there is still to be added to the general precession the quantity $D'L' - DL$. But this part is, on account of the slow change of the obliquity, much smaller than the former.

If we denote by Π the longitude of the ascending node of the true on the fixed ecliptic (that is, the point of intersection of the two great circles, setting out from which the true ecliptic has a north latitude above the second), and reckon this angle from the fixed vernal equinox of the year 1750, then we have, the longitudes being reckoned in the direction from B to D, and E being the descending node of the true ecliptic on the fixed,

$$DE = 180^\circ - \Pi, \quad BE = 180^\circ - \Pi - l_1.$$

Moreover, if the general precession CD' be denoted by l,

$$EC = 180^\circ - \Pi - l.$$

If also we denote by π the angle BEC, that is, the inclination of the true ecliptic to the fixed, we have by Napier's analogies in the triangle BEC

$$\tan\frac{l_1 - l}{2}\cos\frac{\epsilon - \epsilon_0}{2} = \tan\frac{a}{2}\cos\frac{\epsilon + \epsilon_0}{2},$$

$$\tan\frac{\pi}{2}\sin\left(\Pi + \frac{l_1 + l}{2}\right) = \sin\frac{l_1 - l}{2}\tan\frac{\epsilon + \epsilon_0}{2},$$

$$\tan\frac{\pi}{2}\cos\left(\Pi + \frac{l_1 + l}{2}\right) = \cos\frac{l_1 - l}{2}\tan\frac{\epsilon - \epsilon_0}{2}.$$

By these equations l, π, and Π can now be developed in series which proceed according to powers of the time t. The first equation gives

$$\tan\frac{l_1 - l}{2} = \tan\frac{a}{2} \cdot \frac{\cos\frac{\epsilon+\epsilon_0}{2}}{\cos\frac{\epsilon-\epsilon_0}{2}};$$

or, if we put $\epsilon_0 + \frac{\epsilon-\epsilon_0}{2}$ for $\frac{\epsilon+\epsilon_0}{2}$ and replace the sine and tangent of the small angles $l_1 - l$, a, and $\epsilon - \epsilon_0$ by the arcs,

$$l = l_1 - a\cos\epsilon_0 + \frac{\frac{a}{2}(\epsilon-\epsilon_0)\sin\epsilon_0}{206265} \quad\ldots\ldots\ldots\ldots\ (a).$$

In addition,

$$\tan\left(\Pi + \frac{l_1 + l}{2}\right) = \tan\frac{a}{2} \cdot \frac{\sin\frac{\epsilon+\epsilon_0}{2}}{\sin\frac{\epsilon-\epsilon_0}{2}};$$

or, by proceeding as before,

$$\tan\left(\Pi + \frac{l_1 + l}{2}\right) = \frac{a\sin\epsilon_0}{\epsilon-\epsilon_0} + \frac{\frac{a}{2}\cos\epsilon_0}{206265} \quad\ldots\ldots\ldots\ldots\ (b).$$

Finally,

$$\tan^2\frac{\pi}{2} = \left(\tan^2\frac{l_1 - l}{2} - \tan^2\frac{\epsilon+\epsilon_0}{2} + \tan^2\frac{\epsilon-\epsilon_0}{2}\right) \times \cos^2\frac{l_1 - l}{2}.$$

If we substitute here for $\tan\frac{l_1 - l}{2}$ the value found above, we obtain (putting again $\epsilon_0 + \frac{\epsilon-\epsilon_0}{2}$ instead of $\frac{\epsilon+\epsilon_0}{2}$ and replacing the sines of small angles by the arcs, and the cosines of the same by unity)

$$\pi^2 = a^2\sin^2\epsilon_0 + (\epsilon-\epsilon_0)^2 + \frac{a^2\sin\epsilon_0\cos\epsilon_0(\epsilon-\epsilon_0)}{206265} \quad\ldots\ldots\ (c).$$

Putting now in (a), (b), and (c), instead of l_1, a, and $\epsilon - \epsilon_0$, their expressions, which are of the form

$$\lambda t + \lambda' t^2, \quad \alpha t + \alpha' t^2, \quad \eta t + \eta' t^2,$$

we easily obtain

$$l = (\lambda - \alpha\cos\epsilon_0)\,t + \left(\lambda' - \alpha'\cos\epsilon_0 + \frac{\frac{\alpha}{2}.\eta\sin\epsilon_0}{206265}\right)t^2,$$

$$\Pi + \frac{l_1 + l}{2} = \tan^{-1}\frac{\alpha\sin\epsilon_0}{\eta}$$

$$+\,t\left(\frac{\alpha'\eta\sin\epsilon_0 - \alpha\eta'\sin\epsilon_0}{\eta^2}.\,206265 + \frac{\alpha}{2}\cos\epsilon_0\right)\times\sec^2\Pi,$$

$$\pi = t\,.\,\sqrt{(\alpha^2\sin^2\epsilon_0 + \eta^2)} + \frac{t^2}{\pi}\left(\alpha\alpha'\sin^2\epsilon_0 + \eta\eta' + \frac{\frac{1}{2}\alpha^2\eta\sin\epsilon_0\cos\epsilon_0}{206265}\right);$$

or, if we substitute for λ, λ', α, α', and η, η' the values previously given*,

$$l = 50''\!\cdot\!21129\,t + 0''\!\cdot\!0001221483\,t^2,$$

$$\frac{dl}{dt} = +\,50''\!\cdot\!21129 + 0''\!\cdot\!0002442966\,t,$$

$$\pi = +\,0''\!\cdot\!48892\,t - 0''\!\cdot\!0000030715\,t^2,$$

$$\frac{d\pi}{dt} = +\,0''\!\cdot\!48892 - 0''\!\cdot\!000006143\,t,$$

$$\Pi = 171^\circ.\,36'.\,10'' - 5''\!\cdot\!21\,t.$$

6. After obtaining a knowledge of the mutual changes of the planes to which the places of the stars are referred, it is easy to determine the resulting changes of the places of the stars themselves. Denoting by λ and β the longitude and latitude of a star referred to the true ecliptic for the epoch $1750 + t$, the co-ordinates of the star in relation to this fundamental plane, taking as zero of longitude the ascending node of the true ecliptic on the fixed ecliptic, are

$$\cos\beta\cos(\lambda - \Pi - l),\quad \cos\beta\sin(\lambda - \Pi - l),\quad \text{and}\quad \sin\beta.$$

If then L and B be the longitude and latitude of the star, referred to the fixed ecliptic for 1750, the three co-ordinates

* For η and η' the numerical values are to be taken from the following equation:

$$\epsilon - \epsilon_0 = -\,t\,.\,0'',48368 - t^2\,.\,0'',00001256528.$$

referred to this fundamental plane and reckoned from the same zero are

$$\cos B \cos (L - \Pi), \quad \cos B \sin (L - \Pi), \quad \text{and} \quad \sin B.$$

Since now the fundamental planes of the two systems of co-ordinates make with each other the angle π, we obtain by means of formula (2) of the Introduction

$$\left.\begin{aligned} \cos \beta \cos (\lambda - \Pi - l) &= \cos B \cos (L - \Pi) \\ \cos \beta \sin (\lambda - \Pi - l) &= \cos B \sin (L - \Pi) \cos \pi + \sin B \sin \pi \\ \sin \beta &= -\cos B \sin (L - \Pi) \sin \pi + \sin B \cos \pi \end{aligned}\right\} (A).$$

Differentiating these equations, considering L and B as constant, we obtain by means of the differential formula (13) of the Introduction

$$\begin{aligned} d(\lambda - \Pi - l) = d\Pi - \pi \tan \beta \sin (\lambda - \Pi - l)\, d\Pi \\ + \tan \beta \cos (\lambda - \Pi - l)\, d\pi, \end{aligned}$$

$$d\beta = -\pi \cos (\lambda - \Pi - l)\, d\Pi - \sin (\lambda - \Pi - l)\, d\pi.$$

Whence we obtain, after dividing by dt and putting $t\frac{d\pi}{dt}$ instead of π in the coefficient of $d\Pi$, for the annual changes of longitude and latitude of the star, the following formulæ:

$$\frac{d\lambda}{dt} = + \tan \beta \cos \left(\lambda - \Pi - l - \frac{d\Pi}{dt} \cdot t\right) \frac{d\pi}{dt},$$

$$\frac{d\beta}{dt} = - \sin \left(\lambda - \Pi - l - \frac{d\Pi}{dt} \cdot t\right) \frac{d\pi}{dt};$$

or, if we make

$$\Pi + t \cdot \frac{d\Pi}{dt} - l = 171^\circ.36'.10'' + t.39''\!\cdot\!79 = M,$$

$$\left.\begin{aligned} \frac{d\lambda}{dt} &= \frac{dl}{dt} + \tan \beta \cos (\lambda - M) \frac{d\pi}{dt} \\ \frac{d\beta}{dt} &= - \sin (\lambda - M) \frac{d\pi}{dt} \end{aligned}\right\} \ldots\ldots\ldots\ldots (B),$$

where the numerical values for $\frac{dl}{dt}$ and $\frac{d\pi}{dt}$ are given in the preceding No.

Denoting again by L and B the longitude and latitude of a star, referred to the fixed ecliptic and the equinox of 1750, then this longitude, reckoned from the intersection of the equator for the year $1750+t$ with the fixed ecliptic for 1750, will be equal to $L+l_1$ where l_1 is the amount of the lunisolar precession in the interval from 1750 to $1750+t$.

The co-ordinates of the star referred to the plane of the ecliptic for 1750 and the above-mentioned point of intersection will thus be

$$\cos B \cos (L+l_1), \quad \cos B \sin (L+l_1), \quad \text{and} \quad \sin B.$$

Denoting then by α and δ the right ascension and declination of the star, referred to the equator and the true equinox for the year $1750+t$, the right ascension reckoned from the above-mentioned point of intersection will be $\alpha+a$. We have thus for the co-ordinates of the star, referred to the plane of the true equator and the assumed point of intersection,

$$\cos \delta \cos (\alpha+a), \quad \cos \delta \sin (\alpha+a), \quad \text{and} \quad \sin \delta.$$

Since the two systems of co-ordinates make with each other the angle ϵ_0, we obtain by formula (1) of the Introduction

$$\left.\begin{aligned} \cos \delta \cos (\alpha+a) &= \cos B \cos (L+l_1) \\ \cos \delta \sin (\alpha+a) &= \cos B \sin (L+l_1) \cos \epsilon_0 - \sin B \sin \epsilon_0 \\ \sin \delta &= \cos B \sin (L+l_1) \sin \epsilon_0 + \sin B \cos \epsilon_0 \end{aligned}\right\} \quad (C).$$

Differentiating these formulæ again, considering L and B as constant, we obtain by the differential formulæ (13) of the Introduction

$$d(\alpha+a) = [\cos \epsilon_0 + \sin \epsilon_0 \tan \delta \sin (\alpha+a)]\, dl_1 - \cos (\alpha+a) \tan \delta \, d\epsilon_0,$$

$$d\delta = \cos (\alpha+a) \sin \epsilon_0 \, dl_1 + \sin (\alpha+a)\, d\epsilon_0.$$

We have then, for the annual change of right ascension and declination of the stars, the formulæ

$$\frac{d\alpha}{dt} = -\frac{da}{dt} + [\cos \epsilon_0 + \sin \epsilon_0 \tan \delta \sin \alpha] \frac{dl_1}{dt}$$

$$+ \left(a \sin \epsilon_0 \frac{dl_1}{dt} - \frac{d\epsilon_0}{dt}\right) \tan \delta \cos \alpha,$$

$$\frac{d\delta}{dt} = \cos\alpha \sin\epsilon_0 \frac{dl_1}{dt} - \left(a \sin\epsilon_0 \frac{dl_1}{dt} - \frac{d\epsilon_0}{dt}\right) \sin\alpha,$$

or, neglecting the exceedingly small second terms of each equation*,

$$\frac{d\alpha}{dt} = -\frac{da}{dt} + [\cos\epsilon_0 + \sin\epsilon_0 \tan\delta \sin\alpha] \frac{dl_1}{dt},$$

$$\frac{d\delta}{dt} = \cos\alpha \sin\epsilon_0 \,.\, \frac{dl_1}{dt}.$$

And, putting

$$\cos\epsilon_0 \frac{dl_1}{dt} - \frac{da}{dt} = m,$$

$$\text{and } \sin\epsilon_0 \frac{dl_1}{dt} = n,$$

we obtain finally

$$\left.\begin{aligned} \frac{d\alpha}{dt} &= m + n \tan\delta \sin\alpha \\ \frac{d\delta}{dt} &= n \cos\alpha \end{aligned}\right\} \ldots\ldots\ldots\ldots\ldots\ldots (D);$$

and, for the numerical values of m and n, if we substitute the values of ϵ_0, $\frac{dl_1}{dt}$, and $\frac{da}{dt}$,

$$m = 46''{\cdot}02824 + 0''{\cdot}0003086450\, t,$$

$$n = 20''{\cdot}06442 - 0''{\cdot}0000970204\, t.$$

Now to obtain the amount of the precession in longitude and latitude or in right ascension and declination in the interval from $1750 + t$ to $1750 + t'$, it is necessary to take the integrals of the equations (B) or (D) between the limits t and t'. In the meanwhile we may find this amount also to terms of the second order inclusive from the differential coefficients for the time $\frac{t+t'}{2}$ and the interval. If, for instance, $f(t)$ and $f(t')$ be two functions,

* The numerical value of the coefficient $a \sin\epsilon_0 \frac{dl_1}{dt} - \frac{d\epsilon_0}{dt}$, is $-0'',0000022472\, t$.

whose difference $f(t')-f(t)$ is required, as in the present case is wanted the amount of precession in the time $t'-t$, we put

$$\frac{1}{2}(t'+t)=x,$$

$$\frac{1}{2}(t'-t)=\Delta x.$$

Then

$$f(t)=f(x-\Delta x)=f(x)-\Delta x f'(x)+\frac{1}{2}(\Delta x)^2 f''(x),$$

and

$$f(t')=f(x+\Delta x)=f(x)+\Delta x f'(x)+\frac{1}{2}(\Delta x)^2 f''(x);$$

where $f'(x)$ and $f''(x)$ denote the first and second differential coefficients of $f(x)$. Hence we obtain

$$f(t')-f(t)=2\Delta x f'(x)=(t'-t)f'\left(\frac{t'+t}{2}\right).$$

Thus to obtain the precession for an interval $t'-t$, it is only necessary to compute the differential coefficient corresponding to the arithmetical mean of the times and to multiply it by the interval of time. By this means the terms of the second order are taken into account.

If, for example, there be required the precession in longitude and latitude in the time from 1750 to 1850 for a star, whose place for 1750 is

$$\lambda=210^\circ.0'; \quad \beta=+34^\circ.0',$$

then, for 1800, we have for the values of $\frac{dl}{dt}$, $\frac{d\pi}{dt}$, and M,

$$\frac{dl}{dt}=50''\cdot 22350, \quad \frac{d\pi}{dt}=0''\cdot 48861, \text{ and } M=172^\circ.9'.20''.$$

Moreover, we obtain for 1800, by reckoning approximately the precession from 1750 to 1800,

$$\lambda=210^\circ.42',1, \quad \beta=+33^\circ.59',8;$$

and, therefore, by formula (B), for 1800,

$$\frac{d\lambda}{dt}=+50'',48122, \quad \frac{d\beta}{dt}=-0'',30447,$$

and thus, the amount of precession from 1750 to 1850, is

$$\text{in longitude} = +\,1^{\circ}.24'.8'',12,$$

$$\text{and in latitude} = - \qquad 30'',45.$$

In like manner if the amount of precession in right ascension and declination be required from 1750 to 1850, for a star whose right ascension and declination for 1750 are

$$\alpha = 220^{\circ}.1'.24'', \quad \delta = +\,20^{\circ}.21'.15'',$$

then, for 1800, we have

$$m = 46'',04367, \quad n = 20'',05957,$$

and, for the approximate place of the star, for this time,

$$\alpha = 220^{\circ}.35',8, \quad \delta = +\,20^{\circ}.8',6\,;$$

and, from these data, we obtain by formula (D),

$$\begin{array}{rlcrl} \log \tan \delta = & 9\cdot56444 & & \log \cos \alpha = & -\,9\cdot88042 \\ \log \sin \alpha = & -\,9\cdot81340 & & \log n = & 1\cdot30232 \\ \log n = & 1\cdot30232 & & & -\,1\cdot18274 \\ & -\,0\cdot68016 & & \dfrac{d\delta}{dt} = & -\,15'',2314 \\ n \sin \alpha \tan \delta = & -\,4'',78806 & & & \\ m = & +\,46\cdot04367 & & & \\ \dfrac{d\alpha}{dt} = & +\,41\cdot25561 & & & \end{array}$$

and therefore the amount of precession from 1750 to 1850 in right ascension $= +\,1^{\circ}.8'.45'',56$, and in declination $= -\,25'.23'',14$.

7. The differential formulæ given above do not serve, when it is required to compute the precession for times very far distant from each other, or for stars which are very near the pole. In these cases the exact formulæ must be employed.

Let λ and β be the longitude and latitude of a star, referred to the ecliptic and the equinox for the time $1750 + t$, then we obtain the longitude and latitude L and B, referred to the fixed ecliptic for 1750, by the following equations, which follow immediately from the equations (A) of No. 6:

$$\cos B \cos (L - \Pi) = \cos \beta \cos (\lambda - \Pi - l),$$
$$\cos B \sin (L - \Pi) = \cos \beta \sin (\lambda - \Pi - l) \cos \pi - \sin \beta \sin \pi,$$
$$\sin B = \cos \beta \sin (\lambda - \Pi - l) \sin \pi + \sin \beta \cos \pi.$$

If then the longitude and latitude λ' and β', referred to the ecliptic and equinox for the epoch $1750 + t'$ be required, these quantities are obtained from L and B by the following equations, if we denote the values of the quantities Π, π, and l for the time t' by Π', π', and l',

$$\cos \beta' \cos (\lambda' - \Pi' - l') = \cos B \cos (L - \Pi'),$$
$$\cos \beta' \sin (\lambda' - \Pi' - l') = \cos B \sin (L - \Pi') \cos \pi' + \sin B \sin \pi',$$
$$\sin \beta' = - \cos B \sin (L - \Pi') \sin \pi' + \sin B \cos \pi'.$$

By eliminating B and L from these equations we obtain immediately λ' and β' expressed in terms of λ and β, and of l, Π, and π for the epochs t and t'. These formulæ will however be seldom employed, since, for longitudes and latitudes, the differential formulæ previously given, on account of the smallness of the square of π, serve for very great intervals of time. Thus, in the preceding example, the error of the differential formulæ amounts to only 0″,02. For right ascension and declination the rigorous formulæ will be precisely similar. If the right ascension and declination of a star be α and δ for the epoch $1750 + t$, then we obtain the longitude and latitude L and B referred to the fixed ecliptic for 1750 by means of the equations*,

$$\cos B \cos (L + l_1) = \cos \delta \cos (\alpha + a),$$
$$\cos B \sin (L + l_1) = \cos \delta \sin (\alpha + a) \cos \epsilon_0 + \sin \delta \sin \epsilon_0,$$
$$\sin B = - \cos \delta \sin (\alpha + a) \sin \epsilon_0 + \sin \delta \cos \epsilon_0.$$

If the right ascension and declination α' and δ' for the epoch $1750 + t'$ be required, these are obtained from L and B, denoting the values of l_1, a and ϵ_0 for the time t' by l_1', a' and ϵ_0', by means of the following equations:

$$\cos \delta' \cos (\alpha' + a') = \cos B \cos (L + l_1'),$$
$$\cos \delta' \sin (\alpha' + a') = \cos B \sin (L + l_1') \cos \epsilon_0' - \sin B \sin \epsilon_0',$$
$$\sin \delta' = \cos B \sin (L + l_1') \sin \epsilon_0' + \sin B \cos \epsilon_0'.$$

* These equations are easily found from equations (C) in No. 6, by consideration of the spherical triangle between the star, the pole of the ecliptic for 1750, and the pole of the equator for $1750 + t$.

By eliminating now B and L from the two systems of equations we obtain, (since

$$\cos B \sin L = -\cos\delta \cos(\alpha + a) \sin l_1 + \cos\delta \sin(\alpha + a) \cos\epsilon \cos l_1 + \sin\delta \sin\epsilon \cos l_1,$$

$$\cos B \cos L = \cos\delta \cos(\alpha + a) \cos l_1 + \cos\delta \sin(\alpha + a) \cos\epsilon \sin l_1 + \sin\delta \sin\epsilon \sin l_1,$$

$$\sin B = -\cos\delta \cos(\alpha + a) \sin\epsilon + \sin\delta \cos\epsilon,)$$

as is easily seen, the following equations:

$$\cos\delta' \cos(\alpha' + a') = \cos\delta \cos(\alpha + a) \cos(l_1' - l_1) - \cos\delta \sin(\alpha + a) \sin(l_1' - l_1) \cos\epsilon_0 - \sin\delta \sin(l_1' - l_1) \sin\epsilon_0,$$

$$\cos\delta' \sin(\alpha' + a') = \cos\delta \cos(\alpha + a) \sin(l_1' - l_1) \cos\epsilon_0' + \cos\delta \sin(\alpha + a) [\cos(l_1' - l_1) \cos\epsilon_0 \cos\epsilon_0' + \sin\epsilon_0 \sin\epsilon_0'] + \sin\delta [\cos(l_1' - l_1) \sin\epsilon_0 \cos\epsilon_0' - \cos\epsilon_0 \sin\epsilon_0']$$

$$\sin\delta' = \cos\delta \cos(\alpha + a) \sin(l_1' - l_1) \sin\epsilon_0' + \cos\delta \sin(\alpha + a) [\cos(l_1' - l_1) \cos\epsilon_0 \sin\epsilon_0' - \sin\epsilon_0 \cos\epsilon_0'] + \sin\delta [\cos(l_1' - l_1) \sin\epsilon_0 \sin\epsilon_0' + \cos\epsilon_0 \cos\epsilon_0'].$$

Imagine now a spherical triangle, whose three sides are

$$l_1' - l_1, \quad 90^\circ - z, \quad \text{and } 90^\circ + z',$$

and whose three opposite angles are respectively Θ, ϵ_0', and $180^\circ - \epsilon_0$, then the coefficients of the preceding equations which contain $l_1' - l_1$, ϵ_0, and ϵ_0', are expressed by Θ, z, and z', and we consequently find

$$\cos\delta' \cos(\alpha' + a') = \cos\delta \cos(\alpha + a) [\cos\Theta \cos z \cos z' - \sin z \sin z'] + \cos\delta \sin(\alpha + a) [\cos\Theta \sin z \cos z' + \cos z \sin z'] - \sin\delta \sin\Theta \cos z'$$

$$\cos\delta' \sin(\alpha' + a') = \cos\delta \cos(\alpha + a) [\cos\Theta \cos z \sin z' + \sin z \cos z'] - \cos\delta \sin(\alpha + a) [\cos\Theta \sin z \sin z' - \cos z \cos z'] - \sin\delta \sin\Theta \sin z'$$

$$\sin\delta' = \cos\delta \cos(\alpha + a) \sin\Theta \cos z - \cos\delta \sin(\alpha + a) \sin\Theta \sin z + \sin\delta \cos\Theta.$$

If we multiply the first of these equations by $\sin z'$, the second by $\cos z'$, and add them together, and again multiply the first by $\cos z'$ and the second by $\sin z'$, and add the products in the same way, we obtain

$$\left.\begin{aligned}\cos\delta'\sin(\alpha'+a'-z') &= \cos\delta\sin(\alpha+a+z)\\ \cos\delta'\cos(\alpha'+a'-z') &= \cos\delta\cos(\alpha+a+z)\cos\Theta-\sin\delta\sin\Theta\\ \sin\delta' &= \cos\delta\cos(\alpha+a+z).\sin\Theta+\sin\delta\cos\Theta\end{aligned}\right\}..(a).$$

These formulæ serve immediately for expressing α' and δ' in terms of α, δ, a, a' and the auxiliary quantities z, z', and Θ. Finally we find, by employing upon the above-mentioned spherical triangle the Gaussian formulæ,

$$\sin\frac{1}{2}\Theta\cos\frac{1}{2}(z'-z)=\sin\frac{1}{2}(l_1'-l_1)\sin\frac{1}{2}(\epsilon_0'+\epsilon_0),$$

$$\sin\frac{1}{2}\Theta\sin\frac{1}{2}(z'-z)=\cos\frac{1}{2}(l_1'-l_1)\sin\frac{1}{2}(\epsilon_0'-\epsilon_0)$$

$$\cos\frac{1}{2}\Theta\sin\frac{1}{2}(z'+z)=\sin\frac{1}{2}(l_1'-l_1)\cos\frac{1}{2}(\epsilon_0'+\epsilon_0),$$

$$\cos\frac{1}{2}\Theta\cos\frac{1}{2}(z'+z)=\cos\frac{1}{2}(l_1'-l_1)\cos\frac{1}{2}(\epsilon_0'-\epsilon_0).$$

Now here it is always allowable to replace $\sin\frac{1}{2}(z'-z)$ and $\sin\frac{1}{2}(\epsilon_0'-\epsilon_0)$ by the arcs, and to assume the corresponding cosines to be equal to 1, so that for the computation of the three auxiliary quantities we obtain the following simple formulæ:

$$\left.\begin{aligned}\tan\frac{1}{2}(z'+z) &= \cos\frac{1}{2}(\epsilon_0'+\epsilon_0)\tan\frac{1}{2}(l_1'-l_1)\\ \frac{1}{2}(z'-z) &= \frac{1}{2}(\epsilon_0'-\epsilon_0)\,\frac{\operatorname{cotan}\frac{1}{2}(l_1'-l_1)}{\sin\frac{1}{2}(\epsilon_0'+\epsilon_0)}\\ \tan\frac{1}{2}\Theta &= \tan\frac{1}{2}(\epsilon_0'+\epsilon_0)\sin\frac{1}{2}(z'+z)\end{aligned}\right\}\ \ldots.\ (A).$$

The formulæ (a) can be arranged for convenient computation by the introduction of an auxiliary angle, or instead of them another system of equations may be made use of, which are obtained in the same manner as before by the Gaussian formulæ. In fact we obtain the formulæ (a) by employing the three fundamental formulæ of spherical trigonometry to a spherical triangle whose three sides are $90^0-\delta'$, $90^0-\delta$, and Θ, and in which the angles $\alpha+a+z$ and $180^0-\alpha'-a'+z'$ are opposite to the first two sides. Employing instead of these the Gaussian formulæ, we obtain, if we denote the third angle by c, and for shortness make

$$\alpha+a+z=A, \text{ and } \alpha'+a'-z'=A',$$

$$\left.\begin{aligned}
\cos\frac{1}{2}(90^0+\delta')\cos\frac{1}{2}(A'+c)&=\cos\frac{1}{2}(90^0+\delta+\Theta)\cos\frac{1}{2}A\\
\cos\frac{1}{2}(90^0+\delta')\sin\frac{1}{2}(A'+c)&=\cos\frac{1}{2}(90^0+\delta-\Theta)\sin\frac{1}{2}A\\
\sin\frac{1}{2}(90^0+\delta')\cos\frac{1}{2}(A'-c)&=\sin\frac{1}{2}(90^0+\delta+\Theta)\cos\frac{1}{2}A\\
\sin\frac{1}{2}(90^0+\delta')\sin\frac{1}{2}(A'-c)&=\sin\frac{1}{2}(90^0+\delta-\Theta)\sin\frac{1}{2}A
\end{aligned}\right\}\ldots(b).$$

We arrive at a greater degree of accuracy still if the quantity A' be not required, but only the difference $A'-A$. We obtain in fact, by multiplying the first of the equations (a) by $\cos A$, and the second by $\sin A$, and subtracting one from the other, and again, by multiplying the first equation by $\sin A$, and the second by $\cos A$ and adding,

$$\cos\delta'\sin(A'-A)=\cos\delta\sin A\sin\Theta\left[\tan\delta+\tan\frac{1}{2}\Theta\cos A\right]$$

$$\cos\delta'\cos(A'-A)=\cos\delta-\cos\delta\cos A\sin\Theta\left[\tan\delta+\tan\frac{1}{2}\Theta\cos A\right];$$

or

$$\tan(A'-A)=\frac{\sin A\sin\Theta\left[\tan\delta+\tan\frac{1}{2}\Theta\cos A\right]}{1-\cos A\sin\Theta\left[\tan\delta+\tan\frac{1}{2}\Theta\cos A\right]},$$

and, by the Gaussian formulæ, we find

$$\cos\frac{1}{2}c\sin\frac{1}{2}(\delta'-\delta)=\sin\frac{1}{2}\Theta\cos\frac{1}{2}(A'+A)$$

$$\cos\frac{1}{2}c\cos\frac{1}{2}(\delta'-\delta)=\cos\frac{1}{2}\Theta\cos\frac{1}{2}(A'-A).$$

Therefore if we put

$$p=\sin\Theta\left[\tan\delta+\tan\frac{1}{2}\Theta\cos A\right]\ldots\ldots\ldots\ldots\ldots(B),$$

we have

$$\left.\begin{aligned}\tan(A'-A)&=\frac{p\sin A}{1-p\cos A}\\ \tan\frac{1}{2}(\delta'-\delta)&=\tan\frac{1}{2}\Theta\,.\,\frac{\cos\frac{1}{2}(A'+A)}{\cos\frac{1}{2}(A'-A)}\end{aligned}\right\}\ldots\ldots(C).$$

The rigorous computation of the right-ascension and declination of a star for the epoch $1750+t'$ from the right ascension and declination of the same for the epoch $1750+t$, is by this means referred to the computation of the formulæ (A), (B), and (C).

Example.

The right ascension and declination of Polaris for the beginning of the year 1755 are

$$\alpha=10^{0}.\,55'.\,44'',955,$$
$$\delta=87^{0}.\,59'.\,41'',12.$$

If we now compute the place, referred to the equator and equinox of 1850, we have

$$l_1=4'.\,11'',8756 \qquad l_1'=1^{0}.\,23'.\,56'',3541,$$
$$a=0'',8897, \qquad a'=15'',2656,$$
$$\epsilon_0=23^{0}.\,28'.\,18'',0002, \qquad \epsilon_0'=23^{0}.\,28'.\,18'',0984.$$

With these data we obtain from formulæ (A)

$$\frac{1}{2}(z'+z)=0^{0}.\,36'.\,34'',314, \qquad \frac{1}{2}(z'-z)=10'',6286,$$

and therefore

$$z=0^{0}.\,36'.\,23'',685,$$
$$z'=0^{0}.\,36'.\,44'',943,$$

and

$$\Theta = 0^\circ.\,31'.\,45'',600,$$

and therefore

$$A = a + a + z = 11^\circ.\,32'.\,9'',530.$$

Computing then by formulæ (B) and (C) the values of $A' - A$ and $\delta' - \delta$, we find

$$\log p = 9{\cdot}4214471,$$

and

$$A' - A = 4^\circ.\,4'.\,17'',710, \qquad \frac{1}{2}(\delta' - \delta) = 0^\circ.\,15'.\,26'',780,$$

and thus

$$A' = 15^\circ.\,36'.\,27'',240,$$

and finally

$$\alpha' = 16^\circ.\,12'.\,56'',917$$

$$\delta' = 88^\circ.\,30'.\,34'',680.$$

8. Since the point of intersection of the equator on the ecliptic retrogrades on the latter by about 50″,2, the pole of the equator will in course of time describe about the pole of the ecliptic a small circle, of which the radius is equal to the obliquity of the ecliptic*. The pole of the equator is therefore constantly coinciding with other points of the apparent sphere of the heavens, or, at different times, different stars will be in the neighbourhood of it. In our time the extreme star in the tail of the Lesser Bear (α Ursæ Minoris) is the nearest to the pole, and is therefore called the pole-star. This star, whose declination is now greater than $88\frac{1}{2}^\circ$, will continually be approaching nearer to the pole, till its right ascension (at present 16°) shall have become equal to 90°. The declination will then have reached its maximum, $89^\circ.\,32'$, and from that time will again diminish, since the precession in declination for stars lying in the second quadrant is negative.

To find now the position of the pole for any time t, consider the spherical triangle between the pole of the ecliptic for a given epoch t_0, and the poles of the equator P and P' for the times t_0 and t. Denoting then the right ascension and declination of

* Strictly speaking, this radius is not constant, but is equal to the actually existing obliquity of the ecliptic.

the pole at the time t referred to the equator and the equinox of the epoch t_0 by α and δ, the obliquities of the ecliptic at the times t_0 and t by ϵ_0 and ϵ, then the side $PP' = 90^\circ - \delta$, $EP = \epsilon_0$, $EP' = \epsilon$, the angle at $P = 90^\circ + \alpha$, and the angle at ϵ is equal to the general precession in the interval $t - t_0$; and we have therefore by the three fundamental equations of spherical trigonometry,

$$\cos\delta \sin\alpha = \sin\epsilon \cos\epsilon_0 \cos l - \cos\epsilon \sin\epsilon_0,$$

$$\cos\delta \cos\alpha = \sin\epsilon \sin l,$$

$$\sin\delta = \sin\epsilon \sin\epsilon_0 \cos l + \cos\epsilon \cos\epsilon_0.$$

Since this computation in general requires no great accuracy, the place of the pole being always required only approximately, and since besides the diminution of the obliquity is only to be looked upon as proportional to the time for short intervals of time, because it has a period of extremely long duration, we may be permitted to make $\epsilon = \epsilon_0$, and we then obtain the simple equations:

$$\tan\alpha = -\cos\epsilon_0 \tan\frac{1}{2} l,$$

and

$$\cos\delta = \frac{\sin\epsilon_0 \sin l}{\cos\alpha}.$$

Although α is here found by means of the tangent, its value is still found without any ambiguity, since it must at the same time fulfil the condition, that $\cos\alpha$ and $\sin l$ must have the same sign.

If for example the position of the pole were required for the year 14000 referred to the equinox of 1850, then for the interval 12150 years the general precession will amount to about 174°, and therefore we shall have

$$\alpha = 273^\circ.16', \text{ and } \delta = +43^\circ.7'.$$

This is very near the place of α Lyræ whose right ascension and declination for 1850 are

$$\alpha = 277^\circ.58', \text{ and } \delta = +38^\circ.39'.$$

Thus in the year 14000 this star will have pretensions to the name of the Pole-star.

On account of the changes of declination of the stars through the precession, in course of time some will come above the horizon of a place which before were never visible there, and others which are now, for example, visible at a place in the northern hemisphere of the globe, will on the contrary obtain so southern a declination, that they will no longer rise above the horizon of the place. In like manner will stars which at the present time are always above the horizon of the place, begin to have their risings and settings, while, on the contrary, other stars will attain so large a north declination that they even at their lower culmination will remain above the horizon. The aspect of the heavens at any place on the earth will thus be remarkably changed by means of the precession.

In the remark appended to No. 16 of the second section, the sidereal year, or the sidereal period of revolution of the sun, that is the time required by the sun to pass over 360° of the sphere of the heavens, or the time in which it again returns to the same fixed star, was estimated at 365 days, 6 hours, 9 minutes, and 10·7496 seconds, or at 365·25637 mean days. Since now the equinoctial point moves backwards, that is contrary to the motion of the sun, the tropical year, that is the time required by the sun to return to the same equinox, will be shorter than the sidereal year by the time taken by the sun to describe the small arc which is equal to the annual precession. Now, for the year 1800, $l = 50'',2235$, and since the mean daily motion of the sun amounts to $59'.8'',33$, we obtain for this time 0·01415 of a day, and for the length of the tropical year 365·24222 days. But since the precession is variable and its annual increase amounts to $0''\cdot0002442966$, the tropical year is also variable, and the annual change of the same is equal to 0,000000068848 of a day. If the decimal part be expressed in hours, minutes, and seconds, we thus obtain for the length of the tropical year

$$365^d.5^h.48^m.47^s,8091 - 0^s,00595\,(t - 1800).$$

9. The lunisolar precession contains only the terms proportional to the time in the motion of the equator on the fixed ecliptic which are produced by the attraction of the sun and moon on the spheroidal earth. But theory teaches that the complete expression of this motion contains, besides those times, others

periodical, which depend on the places of the sun and moon, but especially on the longitude of the moon's node (that is, the longitude towards which the line of intersection of the plane of the moon's orbit and the ecliptic is directed)*. This periodical part in the motion of the equator on the fixed ecliptic is denoted by the term *Nutation*, since it is in a manner produced by a periodical oscillation of the earth's axis about its mean position, and the periodical motion of the point of intersection is called the *nutation in longitude*, while on the other hand, the periodical part of the change of inclination is called the *nutation of the obliquity of the ecliptic*. The point in which the equator and the ecliptic at any time really intersect each other is called the *true equinox* for that time, and on the other hand the point of intersection cleared of the nutation is called the *mean equinox*. In the same manner by the *true obliquity of the ecliptic* is meant that inclination of the ecliptic to the equator, which, on account of the secular change and the nutation, really exists, while on the other hand by the *mean obliquity* is meant the inclination cleared of the nutation.

The expressions for the change of longitude and obliquity of the ecliptic, $\Delta\lambda$ and $\Delta\epsilon$, are at the present time according to Bessel,

$$\left.\begin{aligned}\Delta\lambda &= -\,16'',78332\ \sin\Omega + 0'',20209\sin 2\Omega\\ &\quad -\,1'',33589\sin 2\odot - 0'',20128\sin 2☽\\ &\text{and}\\ \Delta\epsilon &= +\,8'',97707\cos\Omega - 0''{\cdot}08773\cos 2\Omega\\ &\quad +\,0'',57990\cos 2\odot + 0'',08738\cos 2☽,\end{aligned}\right\}\ (a),$$

where Ω denotes the longitude of the ascending node of the moon's orbit on the ecliptic, and $\odot$ and ☽ denote the longitudes of the sun and moon. Now to compute the amount of the nutation for right ascension and declination, we first obtain, if we denote by α and δ the mean right ascension and declination, the mean longitude and latitude by the formulæ:

$$\begin{aligned}\cos\beta\cos\lambda &= \cos\delta\cos\alpha,\\ \cos\beta\sin\lambda &= \cos\delta\sin\alpha\cos\epsilon + \sin\delta\sin\epsilon,\\ \sin\beta &= -\cos\delta\sin\alpha\sin\epsilon + \sin\delta\cos\epsilon.\end{aligned}$$

* This motion of the moon's node is very rapid, since it amounts to 360° in about 19 years.

Augmenting then the longitudes thus found by the nutation $\Delta\lambda$, and the obliquity of the ecliptic by $\Delta\epsilon$, we find the apparent right ascension and declination α' and δ' by the equations:

$$\cos\delta'\cos\alpha' = \cos\beta\cos(\lambda+\Delta\lambda),$$
$$\cos\delta'\sin\alpha' = \cos\beta\sin(\lambda+\Delta\lambda)\cos(\epsilon+\Delta\epsilon) - \sin\beta\sin(\epsilon+\Delta\epsilon),$$
$$\sin\delta' = \cos\beta\sin(\lambda+\Delta\lambda)\sin(\epsilon+\Delta\epsilon) + \sin\beta\cos(\epsilon+\Delta\epsilon).$$

But, since the variations $\Delta\lambda$ and $\Delta\epsilon$ are but small, differential formulæ will be sufficient for the purpose. We shall have, namely,

$$\left.\begin{aligned} \alpha'-\alpha &= \left(\frac{d\alpha}{d\lambda}\right)\Delta\lambda + \left(\frac{d\alpha}{d\epsilon}\right)\Delta\epsilon + \frac{1}{2}\left(\frac{d^2\alpha}{d\lambda^2}\right)(\Delta\lambda)^2 + \left(\frac{d^2\alpha}{d\lambda\, d\epsilon}\right)\Delta\lambda\,\Delta\epsilon \\ &\quad + \frac{1}{2}\left(\frac{d^2\alpha}{d\epsilon^2}\right)(\Delta\epsilon)^2 + \&c. \\ \text{and} & \\ \delta'-\delta &= \left(\frac{d\delta}{d\lambda}\right)\Delta\lambda + \left(\frac{d\delta}{d\epsilon}\right)\Delta\epsilon + \frac{1}{2}\left(\frac{\Delta^2\delta}{d\lambda^2}\right)(\Delta\lambda)^2 + \left(\frac{d^2\delta}{d\lambda\, d\epsilon}\right)\Delta\lambda\,\Delta\epsilon \\ &\quad + \frac{1}{2}\left(\frac{d^2\delta}{d\epsilon^2}\right)(\Delta\epsilon)^2 + \&c. \end{aligned}\right\} \quad (b).$$

But, according to the differential formulæ in Section I, No. 10, we have, if for $\cos\beta\sin\eta$ and $\cos\beta\cos\eta$ we put the expressions for them in terms of α and δ,

$$\left(\frac{d\alpha}{d\lambda}\right) = \cos\epsilon + \sin\epsilon\tan\delta\sin\alpha \qquad \left(\frac{d\delta}{d\lambda}\right) = \cos\alpha\sin\epsilon,$$

$$\left(\frac{d\alpha}{d\epsilon}\right) = -\cos\alpha\tan\delta \qquad \left(\frac{d\delta}{d\epsilon}\right) = \sin\alpha,$$

from which we obtain by differentiation:

$$\frac{d^2\alpha}{d\lambda^2} = \sin^2\epsilon\,[\tfrac{1}{2}\sin 2\alpha + \text{cotan}\,\epsilon\cos\alpha\tan\delta + \sin 2\alpha\tan^2\delta],$$

$$\frac{d^2\alpha}{d\lambda\, d\epsilon} = -\sin\epsilon\,[\cos^2\alpha - \text{cotan}\,\epsilon\tan\delta\sin\alpha + \tan^2\delta\cos 2\alpha]$$

$$\frac{d^2\alpha}{d\epsilon^2} = -\,[\tfrac{1}{2}\sin 2\alpha + \sin 2\alpha\tan^2\delta].$$

$$\frac{d^2\delta}{d\lambda^2} = -\sin^2\epsilon\sin\alpha\,[\text{cotan}\,\epsilon + \tan\delta\sin\alpha],$$

$$\frac{d^2\delta}{d\lambda\, d\epsilon} = \sin\epsilon\cos\alpha\left[\cot\!\text{an}\,\epsilon + \sin\alpha\tan\delta\right]$$

$$\frac{d^2\delta}{d\epsilon^2} = -\cos^2\alpha\tan\delta.$$

Substituting these expressions in the equations (*b*), and putting besides for $\Delta\lambda$ and $\Delta\epsilon$ the values previously given from equations (*a*), and for ϵ the mean obliquity for the beginning of the year $1800 = 23^\circ.27'.54''$, we obtain for the terms of the first order,

$$\begin{aligned}
\alpha' - \alpha = {} & -15'',39537\sin\Omega - [6'',68299\sin\Omega\sin\alpha \\
& \qquad\qquad + 8'',97707\cos\Omega\cos\alpha]\tan\delta \\
& + 0'',18538\sin 2\Omega + [0'',08046\sin 2\Omega\sin\alpha \\
& \qquad\qquad + 0'',08773\cos 2\Omega\cos\alpha]\tan\delta \\
& - 1'',22542\sin 2\odot - [0'',53194\sin 2\odot\sin\alpha \\
& \qquad\qquad + 0'',57990\cos 2\odot\cos\alpha]\tan\delta \\
& - 0'',18463\sin 2\text{☽} - [0'',08015\sin 2\text{☽}\sin\alpha \\
& \qquad\qquad + 0'',08738\cos 2\text{☽}\cos\alpha]\tan\delta \qquad (A)
\end{aligned}$$

$$\begin{aligned}
\delta' - \delta = {} & -6'',68299\sin\Omega\cos\alpha + 8'',97707\cos\Omega\sin\alpha \\
& + 0'',08046\sin 2\Omega\cos\alpha - 0'',08773\cos 2\Omega\sin\alpha \\
& - 0'',53194\sin 2\odot\cos\alpha + 0'',57990\cos 2\odot\sin\alpha \\
& - 0'',08015\sin 2\text{☽}\cos\alpha + 0'',08738\cos 2\text{☽}\sin\alpha.
\end{aligned}$$

Of the terms of the second order those only can be significant which arise from the large terms of $\Delta\lambda$ and $\Delta\epsilon$. Putting

$$\Delta\epsilon = 8'',97707\cos\Omega = a\cos\Omega,$$

and

$$-\sin\epsilon\Delta\lambda = 6'',68299\sin\Omega = b\sin\Omega,$$

these terms will be

$$\begin{aligned}
\alpha' - \alpha = {} & \frac{b^2 - a^2}{4}\sin 2\alpha\left[\tan^2\delta + \frac{1}{2}\right] + \frac{b^2}{4}\tan\delta\cos\alpha\,\text{cotan}\,\epsilon \\
& + \left[\frac{1}{2} - \text{cotan}\,\epsilon\sin\alpha\tan\delta + \tan^2\delta\cos 2a + \frac{1}{2}\cos 2\alpha\right]\frac{ab}{2}\sin 2\Omega \\
& - \left[\frac{b^2 + a^2}{4}\tan^2\delta\sin 2\alpha + \frac{b^2}{4}\tan\delta\cos\alpha\,\text{cotan}\,\epsilon \right. \\
& \qquad\qquad \left. + \frac{b^2 + a^2}{8}\sin 2\alpha\right]\cos 2\Omega,
\end{aligned}$$

and

$$\delta' - \delta = -\frac{a^2 \cos^2 \alpha + b^2 \sin^2 \alpha}{4} \tan \delta - \frac{b^2}{4} \sin \alpha \operatorname{cotan} \epsilon$$

$$- (\tan \delta \sin 2\alpha + 2 \operatorname{cotan} \epsilon \cos \alpha) \frac{ab}{4} \sin 2\Omega$$

$$- \left(\frac{a^2 \cos^2 \alpha - b^2 \sin^2 \alpha}{4} \tan \delta - \frac{b^2}{4} \sin \alpha \operatorname{cotan} \epsilon \right) \cos 2\Omega.$$

Of these terms those which are independent of Ω change merely the mean place of the star, and may therefore be neglected. Another portion of the terms, namely,

$$\frac{ab}{4} \sin 2\Omega - \left(\frac{ab}{2} \operatorname{cotan} \epsilon \sin \alpha \sin 2\Omega + \frac{b^2}{4} \cot \epsilon \cos \alpha \cos 2\Omega \right) \tan \delta,$$

and

$$-\frac{ab}{2} \operatorname{cotan} \epsilon \sin 2\Omega \cos \alpha + \frac{b^2}{4} \operatorname{cotan} \epsilon \sin \alpha \cos 2\Omega,$$

are combined with the similar terms multiplied into $\sin 2\Omega$ and $\cos 2\Omega$ of the first order, so that there will be,

in α,

$$+ 0''{\cdot}18545 \sin 2\Omega + (0''{\cdot}08012 \sin 2\Omega \sin \alpha$$
$$+ 0''{\cdot}08761 \cos 2\Omega \cos \alpha) \tan \delta;$$

and in δ,

$$+ 0''{\cdot}08012 \sin 2\Omega \cos \alpha - 0''{\cdot}08761 \cos 2\Omega \sin \alpha. \qquad (B)$$

The yet remaining terms of the second order are then the following:

in right ascension,

$$+ 0''{\cdot}0001454 \left(\tan^2 \delta + \frac{1}{2} \right) \cos 2\alpha \sin 2\Omega,$$

$$- 0''{\cdot}0001518 \left(\tan^2 \delta + \frac{1}{2} \right) \sin 2\alpha \cos 2\Omega;$$

and in declination,

$$- 0''{\cdot}0000727 \tan \delta \sin 2\alpha \sin 2\Omega,$$
$$- (0''{\cdot}0000217 + 0''{\cdot}0000759 \cos 2\alpha) \tan \delta \cos 2\Omega.$$

But since the former terms only attain to the value $0^s,01$ for declinations equal to $88^\circ.10'$, and the others have only the value

0″,01 for declinations equal to 89°. 26′, they may be always safely neglected.

10. For the purpose of more easily computing the nutation in right ascension and declination, tables have been constructed. The terms

$$-15''{\cdot}39537 \sin \Omega = c \text{ and } -1'',22542 \sin 2\odot = g$$

are first tabulated, the arguments of the tables being Ω and 2⊙.

The separate terms for right ascension multiplied into tan δ have always the form

$$a \cos \beta \cos \gamma + b \sin \beta \sin \gamma = A\,(\alpha \cos \beta \cos \gamma + \sin \beta \sin \gamma).$$

But to every expression of this form can be given the following form,

$$x \cos (\beta - \gamma + y) \ldots\ldots\ldots\ldots\ldots\ldots\ldots (a),$$

by only giving suitable values to the quantities x and y. But by expanding the last expression and comparing it with the former, we obtain for the determination of x and y the equations:

$$A\alpha \cos \beta = x\,(\cos \beta \cos y - \sin \beta \sin y),$$
$$A \sin \beta = x\,(\sin \beta \cos y + \cos \beta \sin y)\,;$$

from whence for x and y we obtain the values

$$x^2 = A^2 \{1 - (1 - \alpha^2) \cos^2 \beta\},$$

and

$$\tan y = \frac{(1-\alpha) \sin \beta \cos \beta}{1 - (1-\alpha) \cos^2 \beta}.$$

Tabulating then the values of x and y, whose argument is β, the expression (a) can be easily computed.

In like manner can tables be constructed for the corresponding terms of the nutation in declination, since these are of the form

$$A\,(-\alpha \cos \beta \sin \gamma + \sin \beta \cos \gamma)\,;$$

and, for every such expression we may always put

$$x \sin (\beta - \gamma + y),$$

where x and y have the same values as before.

Such a table for the nutation, computed by Nicolai, will be found in Warnstorff's Auxiliary Tables, but the fundamental constants are different from those given above, that is to say, from the Constants of Peters*. In these tables will be found, in addition to the term c, the quantities $\log b$ and B, with argument Ω, and there will by this means be obtained the terms depending on $\sin \Omega$ and $\cos \Omega$, which are, for right ascension,

$$c - b \tan \delta \cos (\Omega + B - \alpha),$$

and, for declination,

$$- b \sin (\Omega + B - \alpha).$$

This part of nutation is called *Lunar Nutation.*

A second table gives with argument $2\odot$ the quantities g, F, and $\log f$, by means of which are found the terms depending on $2\odot$, which are, for right ascension,

$$g - f \tan \delta \cos (2\odot + F - \alpha);$$

and for declination,

$$f \sin (2\odot + F - \alpha).$$

This part of nutation is known by the name of *Solar Nutation.*

The terms of nutation depending on the arguments $2☽$ and 2Ω, are then given by the table for solar nutation, if, instead of entering it with $2\odot$, it be once entered with $2\odot$ and a second time with $2\odot + 180^\circ$ (since the last terms have opposite signs); and lastly, if, from the sum of the results corresponding to these two arguments one-sixth part be subtracted, since this is approximately the ratio of its coefficient to that of solar nutation.

11. Since we have by this means obtained a knowledge of the changes of the planes to which the places of the stars are referred, the absolute right ascension of a star can be determined with every accuracy by introducing these changes into the computation. Therefore in the next place it is necessary to free the observations of the sun and the star from nutation. Taking the example previously given in No. 4 of this section the declinations of the sun at the two observations were, for

* Given in his memoir *Numerus Constans Nutationis in Ascensionibus Rectis Stellæ Polaris in Speculâ Dorpatensi annis* 1822 *ad* 1838, *observatis deductus.* Petropoli, 1842.—TRANSLATOR.

$$\text{Mar. 23,}\ D = +1^\circ.6'.54'',$$
$$\text{Sept. 20,}\ D' = +1^\circ.1'.57''.$$

And the right ascensions of the sun and the star, which are known by the previous determination and the observed difference of right ascension, are

$$\text{Mar. 23,}\ A = 2^\circ.34',$$
$$\text{Sept. 20,}\ A' = 177^\circ.37',$$

and

$$\alpha = 112^\circ.35'.$$

Moreover the longitude of the ascending node of the lunar orbit was, at the time of the two observations,

$$\Omega = 207^\circ.21, \quad \Omega' = 197^\circ.45';$$

and the longitudes of the sun

$$\odot = 2^\circ.49', \quad \odot' = 177^\circ.26'.$$

With these values we find:

Nutation for the sun.

Mar. 23, in right ascension $= +0^s{,}48$, in declination $= +2''{,}8$,

Sept. 20, ... $+0^s{,}34$, ... $-2''{,}4$;

and the nutation in right ascension for α Canis Minoris (Procyon), if we assume $\delta = +5^\circ.39'$,

for Mar. 23, $= +0^s{,}47$, and Sep. 20 $= +0^s{,}30$.

If these values for the nutation be applied with changed signs to the observed times of transit (which differ from the right ascensions only by the amount of clock-error) and to the declinations, we obtain those quantities referred to the mean equinox of the days of observation. Finally, account must be taken of the change of the equinox through precession, or all the observed data must be referred to a fixed equinox. If we take as epoch the beginning of the year 1828, we obtain for the precession for the place of the sun,

$$\text{Mar. 23,}\ \Delta A = +0^s{,}71,\ \Delta D = +4''{,}6,$$
$$\text{Sept. 20,}\ \Delta A' = +2^s{,}23,\ \Delta D' = -14''{,}5;$$

and, for the star,

$$\text{Mar. 23,}\ \Delta\alpha = +0^s{,}73, \quad \text{Sep. 20,}\ \Delta\alpha' = +2^s{,}32.$$

Applying these values with changed signs to the observed data, we find, after having taken account of all corrections,

$$T = 0^h.\,11^m.\,12^s{,}62, \qquad T' = 11^h.\,50^m.\,32^s{,}06,$$
$$t = 7\,.\,31\,.\,13{,}00, \qquad t' = 7\,.\,30\,.\,22{,}74,$$
$$D = +1^0.\ 6''.\,54',9, \qquad D' = +1^0.\ 2''.\ 5',6.$$

Finally, we obtain for the obliquity of the ecliptic for both epochs, regard being had to the secular variations and the nutation,

$$\epsilon = 23^0.\,27'.\,33'',9, \qquad \epsilon' = 23^0.\,27'.\,33'',1,$$

and from thence

$$\frac{1}{2}\left[\sin^{-1}\frac{\tan D}{\tan\epsilon} - \sin^{-1}\frac{\tan D'}{\tan\epsilon'}\right] + 6^h = 6^h.\ 0^m.\,22^s{,}25,$$
$$\frac{1}{2}(t-T) + \frac{1}{2}(t'-T') \qquad = 1\,.\,29\,.\,55{,}53\,;$$

and therefore the right ascension of α Canis Minoris referred to the mean equinox of the beginning of the year 1828,

$$\alpha = 7^h.\,30^m.\,17^s{,}78.$$

If now the absolute right ascensions and declinations of the stars be determined at different epochs, we obtain from the comparison of the two positions the amount of the precession in right ascension and declination in the interval, and are able by this means to determine the values of m and n (No. 6) as well as the annual lunisolar precession. But we shall always find that from different stars different values of these constants are obtained, since the stars, besides the apparent motions before treated of, have also proper motions, by virtue of which they will really change their positions in space. Since now these proper motions, as in the case of the precessions, at least for intervals of time not very great, appear to be proportional to the time, we can determine the values of both changes in no other way than by computing the values of the lunisolar precession from a very great number of stars and taking the arithmetical mean of the separate results, assuming thereby that the proper motions, which have different values and in different directions for different stars, will be eliminated in this mean value. The differences which then exhibit themselves in comparison with observations made at the time t', when the place of a star for that epoch is deduced from observations made at the time t by means of the value of the lunisolar precession thus determined, is considered to be the proper motion of the star in right ascension and declination during the time $t'-t$.

On account of the changes of position of the stars through precession and proper motion, catalogues of their places, or star-catalogues, are in all cases serviceable only for one particular epoch. For the purpose then of reducing star-places from one epoch to another more conveniently, it is customary to give for each star in these catalogues the yearly changes in right ascension and declination through precession and proper motion under the title of annual variation and proper motion, and, in addition, the change of the annual variation in a hundred years or the secular variation. If then t_0 be the epoch of the catalogue, the change of place of the star during the time $t-t_0$ is equal to

$$[\text{annual var.} + \text{proper motion} + \frac{t-t_0}{200}\times \text{secular var.})\times(t-t_0).$$

The computation of the secular variation is made by the following formulæ. From No. 6 the annual changes of right ascension and declination are

$$\frac{d\alpha}{dt}=m+n\tan\delta\sin\alpha,$$

$$\frac{d\delta}{dt}=n\cos\alpha.$$

By differentiating these equations, treating all the quantities as variable, and denoting the annual variations of m and n by m' and n', we easily obtain,

$$\frac{d^2\alpha}{dt^2}=n^2\tan^2\delta\sin 2\alpha+n^2\sin 2\alpha+mn\tan\delta\cos\alpha$$
$$+m'+n'\tan\delta\sin\alpha,$$

$$\frac{d^2\delta}{dt^2}=-n^2\sin^2\alpha\tan\delta-mn\sin\alpha+n'\cos\alpha,$$

and, multiplying these second differential coefficients by 100, we obtain the secular variation for right ascension and declination.

12. As has been remarked, it is always assumed that the proper motions of the stars are proportional to the time, and take place in a fixed great circle. These two conditions are not strictly correct, but on account of the extremely slow motions of the stars no sensible error arises from this assumption. But since the fundamental planes to which the places of the stars are referred, are variable, on this account the components of the proper motions in the directions of the polar co-ordinates, which are referred to these planes, are also variable.

The formulæ which, for an assumed equinox, express the polar co-ordinates referred to the equator for the time t' by means of the co-ordinates referred to another equinox for the time t, are, by No. 7,

$$\cos\delta' \sin(\alpha' + a' - z') = \cos\delta \sin(\alpha + a + z),$$
$$\cos\delta' \cos(\alpha' + a' - z') = \cos\delta \cos(\alpha + a + z)\cos\Theta - \sin\delta \sin\Theta,$$
$$\sin\delta' = \cos\delta \cos(\alpha + a + z)\sin\Theta + \sin\delta \cos\Theta,$$

where a denotes the precession produced by the planets during the time $t' - t$, and z, z', and Θ are auxiliary quantities obtained by means of the formulæ (A) of the same No. Since the proper motions are so small, that their squares and products may be neglected, we obtain by the first and third formulæ (13) in No. 9 of the Introduction, remembering that the formulæ above are derived from a triangle of which the sides are $90^\circ - \delta'$, $90^\circ - \delta$, and Θ, and of which the angles are $\alpha + a + z$, $180^\circ - \alpha' - a' + z'$, and c,

$$\Delta\delta' = \cos c\,\Delta\delta - \sin\Theta \sin(\alpha' + a' - z')\,\Delta\alpha,$$
$$\cos\delta'\,\Delta\alpha' = \sin c\,\Delta\delta + \cos\delta \cos c\,\Delta\alpha,$$

or, if $\sin c$ and $\cos c$ be expressed in terms of the other parts of the triangle,

$$\left.\begin{aligned}\Delta\alpha' &= \Delta\alpha\,[\cos\Theta + \sin\Theta \tan\delta' \cos(\alpha' + a' - z')] \\ &\qquad + \frac{\Delta\delta}{\cos\delta}\sin\Theta\,\frac{\sin(\alpha' + a' - z')}{\cos\delta'} \\ \Delta\delta' &= -\Delta\alpha \sin\Theta \sin(\alpha' + a' - z') \\ &\qquad + \frac{\Delta\delta}{\cos\delta}\cos\delta'\,[\cos\Theta + \sin\Theta \tan\delta' \cos(\alpha' + a' - z')]\end{aligned}\right\}\ \ldots (a),$$

and, in the same manner,

$$\left.\begin{aligned}\Delta\alpha &= \Delta\alpha'\left[\cos\Theta - \sin\Theta\tan\delta\cos(\alpha+a+z)\right] \\ &\qquad - \frac{\Delta\delta'}{\cos\delta'}\sin\Theta\frac{\sin(\alpha+a+z)}{\cos\delta} \\ \Delta\delta &= \Delta\alpha'\sin\Theta\sin(\alpha+a+z) \\ &\qquad + \frac{\Delta\delta'}{\cos\delta'}\cos\delta\left[\cos\Theta - \sin\Theta\tan\delta\cos(\alpha+a+z)\right]\end{aligned}\right\} \ldots (b).$$

Example.

The mean right ascension and declination of Polaris for the beginning of the year 1755, is

$$\alpha = 10^\circ.55'.44'',955, \quad \delta = +87^\circ.59'.41'',12.$$

By application of the precession the place of Polaris was in No. 7 computed for 1850, Jan. 1, and found to be

$$\alpha' = 16^\circ.12'.56'',917, \quad \delta' = +88^\circ.30'.34'',680.$$

But in Bessel's *Tabulæ Regiomontanæ* this place is

$$\alpha' = 16^\circ.15'.19'',530, \quad \delta' = +88'.30'.34'',898.$$

This difference arises from the proper motion of Polaris, which thus amounts to $+2'.22'',613$ in right ascension, and to $+0'',218$ in declination in the interval from 1755 to 1850. The annual proper motion of Polaris referred to the equator of 1850 is therefore

$$\Delta\alpha' = +1'',501189, \quad \Delta\delta' = +0'',002295.$$

If we would from this, for example, have the proper motion of Polaris referred to the equator of 1755, it must be computed from the formulæ (*b*).

But we have

$$\Theta = 0^\circ.31'.45'',600,$$
$$\alpha + a + z = 11^\circ.32'.9'',530,$$

and by these numbers we obtain

$$\Delta\alpha = +1'',10836, \quad \Delta\delta = +0'',005063.$$

If now the place of Polaris for 1755 and the proper motion were given, and from these data the place for 1850 were to be computed, we should have

$$95\Delta\alpha = +1'.45'',294,$$
$$95\Delta\delta = + \qquad 0,481,$$

therefore the place for 1755 + the proper motion to 1850, is

$$\alpha = 10^\circ.57'.30'',249,$$

$$\delta = 87^\circ.59'.41'',601,$$

and, with these values, the computation in No. 7 must be repeated, when we should find for α' and δ' the values of the *Tabulæ Regiomontanæ.*

NOTE. On Precession and Nutation compare the preface to the *Tabulæ Regiomontanæ,* page III. et seq.

III. *Mean and Apparent Places of the Fixed Stars.*

13. By means of what has preceded we are now in a condition to compute the mean place of a star for any time whatever, if this be known for any one epoch. In the same way can we from these mean places find the apparent places, or the places which the stars really appear to occupy on the sphere of the heavens, by applying to the mean places the separate corrections, nutation, aberration, refraction, and annual parallax. The last of these corrections is for the fixed stars insensible, (excepting a few stars whose annual parallaxes are however extremely small), and since the refraction is always applied to the observations themselves, there remain only the aberration and nutation, whose values can, as has been before seen, be taken out of tables. Since however these reductions from mean to apparent places and *vice versâ* must be performed in large masses, tables still more convenient have been constructed, which have the time for argument, and enable us to perform with greater facility the reduction from the mean places at the beginning of the year to the apparent places for any particular day of the year. Thus if the mean place of a star be known for any epoch, and it is required to find the apparent place for any given day of another year, the mean place for the beginning of this latter year must first be found by application of the precession, and, when necessary, of the proper motion, and afterwards the reduction to the apparent place for the given day must be taken out of the tables. These tables are given by Bessel.

Let α and δ denote the mean right ascension and declination of a star for the beginning of a year, α' and δ' the apparent right ascension and declination for the time τ, which is reckoned from the beginning of the year and is expressed in parts of it, then

$$\begin{aligned}
\alpha' = \alpha &+ \tau\,[m + n \tan\delta \sin\alpha] + \tau\mu, \text{ precession and proper motion,}\\
&\left.\begin{aligned}
&- [15'',39537 + 6'',68299 \tan\delta \sin\alpha] \sin\Omega\\
&- 8'',97707 \tan\delta \cos\alpha \cos\Omega\\
&+ [0'',18538 + 0'',08046 \tan\delta \sin\alpha] \sin 2\Omega\\
&+ 0'',08773 \tan\delta \cos\alpha \cos 2\Omega\\
&- [1'',22542 + 0'',53194 \tan\delta \sin\alpha] \sin 2\Omega\\
&- 0'',57990 \tan\delta \cos\alpha \cos 2\odot
\end{aligned}\right\}\text{ nutation,}\\
&\left.\begin{aligned}
&- 20'',255 \cos\epsilon \sec\delta \cos\alpha \cos\odot\\
&- 20'',255 \sec\delta \sin\alpha \sin\odot
\end{aligned}\right\}\text{ aberration,}
\end{aligned}$$

$$\begin{aligned}
\text{and } \delta' = \delta &+ \tau n \cos\alpha + \tau\mu' \ldots \text{ precession and proper motion,}\\
&\left.\begin{aligned}
&- 6'',68299 \cos\alpha \sin\Omega + 8'',97707 \sin\alpha \cos\Omega\\
&+ 0'',08046 \cos\alpha \sin 2\Omega - 0'',08773 \sin\alpha \cos 2\Omega\\
&- 0'',53194 \cos\alpha \sin 2\odot + 0'',57990 \sin\alpha \cos 2\odot
\end{aligned}\right\}\text{ nutation,}\\
&\left.\begin{aligned}
&+ 20'',255\,[\sin\alpha \sin\delta \cos\epsilon - \cos\delta \sin\epsilon] \cos\odot\\
&- 20'',255 \cos\alpha \sin\delta \sin\odot
\end{aligned}\right\}\text{ aberration.}
\end{aligned}$$

In these formulæ those terms of the nutation which depend on twice the moon's longitude are not introduced, since they only alter the place of a star by 0'',1, and besides, on account of the rapid motion of the moon, have so short a period, that they are in a great measure eliminated in the mean of several observations.

To tabulate these expressions for $\alpha' - \alpha$ and $\delta' - \delta$ Bessel puts

$$\begin{aligned}
6'',68299 &= ni, & 15'',39537 - mi &= h,\\
0'',08046 &= ni', & 0'',18538 - mi' &= h',\\
0'',53194 &= ni'', & 1'',22542 - mi'' &= h''.
\end{aligned}$$

Then the preceding formulæ can be written thus:

$$\begin{aligned}
\alpha' = \alpha &+ [\tau - i \sin\Omega + i' \sin 2\Omega - i'' \sin 2\Omega] \times [m + n \tan\delta \sin\alpha]\\
&- [8'',97707 \cos\Omega - 0'',08773 \cos 2\Omega + 0'',5799 \cos 2\odot] \times \tan\delta \cos\alpha\\
&- 20'',255 \cos\epsilon \cos\odot \sec\delta \cos\alpha\\
&- 20'',255 \sin\odot \sec\delta \sin\alpha\\
&+ \tau\mu\\
&- h \sin\Omega + h' \sin 2\Omega - h'' \sin 2\odot,
\end{aligned}$$

and

$$\delta' = \delta + [\tau - i \sin \Omega + i' \sin 2\Omega - i'' \sin 2\odot] \times n \cos \alpha$$
$$+ [8'',97707 \cos \Omega - 0'',08773 \cos 2\Omega + 0'',5799 \cos 2\odot] \sin \alpha$$
$$- 20'',255 \cos \epsilon \cos \odot [\tan \epsilon \cos \delta - \sin \delta \sin \alpha]$$
$$- 20'',255 \sin \odot \sin \delta \cos \alpha$$
$$+ \tau\mu'.$$

Let now the following notation be introduced:

$$\left.\begin{array}{ll} A = \tau - i \sin \Omega + i' \sin 2\Omega - i'' \sin 2\odot & \\ B = -8'',97707 \cos \Omega + 0'',08773 \cos 2\Omega - 0'',5799 \cos 2\odot & \\ C = -20'',255 \cos \epsilon \cos \odot & \\ D = -20'',255 \sin \odot & \\ E = -h \sin \Omega + h' \sin 2\Omega - h'' \sin 2\odot & \\ a = m + n \tan \delta \sin \alpha, & a' = n \cos \alpha \\ b = \tan \delta \cos \alpha, & b' = - \sin \alpha \\ c = \sec \delta \cos \alpha, & c' = \tan \epsilon \cos \delta - \sin \delta \sin \alpha \\ d = \sec \delta \sin \alpha, & d' = \sin \delta \cos \alpha \end{array}\right\} \ldots (a);$$

and thus we have the simple expressions:

$$\left.\begin{array}{l} \alpha' = \alpha + Aa + Bb + Cc + Dd + \tau\mu + E \\ \delta' = \delta + Aa' + Bb' + Cc' + Dd' + \tau\mu' \end{array}\right\} \ldots\ldots (b),$$

where the quantities a, b, c, d, a', b', c', d', depend solely on the place of the star and the obliquity of the ecliptic, while, on the contrary, A, B, C, D, depend solely on the time (since $\odot$ and Ω are functions of the time) and so can be tabulated with the time for argument. Bessel has given the values of the logarithms of A, B, C, and D in his *Tabulæ Regiomontanæ*, for every tenth day from 1750 to 1850.

The numerical values of the quantities i and h are

for 1750 $i = 0{\cdot}33308$ $i' = 0{\cdot}00401$ $i'' = 0{\cdot}02651$,
... 1850 $\quad 0{\cdot}33324$ $\quad 0{\cdot}00401$ $\quad 0{\cdot}02652$,

and

for 1750 $h = 0{\cdot}0645$ $h' = 0{\cdot}0008$ $h'' = 0{\cdot}0051$,
... 1850 $\quad 0{\cdot}0468$ $\quad 0{\cdot}0006$ $\quad 0{\cdot}0037$.

It follows therefore that the quantity E never amounts to more than a few hundredths of a second, and therefore, in most cases, may be neglected.

14. The arguments of the tables calculated by Bessel are the days of the year the beginning of which is taken for the instant when the sun's longitude is 280°. These tables therefore serve for that meridian for which the sun, at the commencement of the civil year, has this longitude. But since the sun does not complete a revolution in a whole number of days, but in 365 days and a fraction, these tables for any other year will serve for a different meridian.

Now from the formula (G), in No. 2 of the second section, it follows, that the angle between two meridians on the earth's surface is equal to the difference of the sidereal times at those places, and is thus also equal to the difference of the mean times. Denote then by k, expressed in time, the difference of meridians of Paris and of a place for which the sun at the beginning of the year has the longitude 280°, and consider k to be positive when the place is east of Paris; denote also by d the meridian difference of any other place from Paris considering it positive when the place is west of Paris, then to the time of this latter place, for which the constants A, B, C, D, are required from the tables, must be added the quantity $k + d$, and the tables must be entered with this corrected time. The values of k are given by Bessel in the *Tabulæ Regiomontanæ* for every year from 1750 to 1850.

Now in the tables are found the constants A, B, C, D, for the beginning of the fictitious year, which commences when the longitude of the sun is 280°, and then, for the same time of every tenth sidereal day; they thus apply constantly for $18^h.40^m$ sidereal time of that meridian, for which the sun at the beginning of the year has the longitude aforesaid. If now we desire to take from the tables the value for another sidereal time α, which in this instance we have taken equal to the right ascension of the star, for the purpose of obtaining the apparent place—say for the time of culmination, then there must be added to the argument $k + d$ the quantity

$$\alpha' = \frac{\alpha - 18^h.40^m}{24^h}.$$

Since, moreover, on that day, on which the right ascension of the sun is equal to the right ascension of the star, two culminations of the star must happen, we must at this time add a unit to the datum of the day, so that the complete argument will be equal to the datum + the quantity

$$k + d + \alpha' + i,$$

where i is equal to 0 from the beginning of the year to the time when the right ascension of the sun is equal to α, but afterwards is equal to + 1.

The day denoted in the tables by Jan. 0 is that, at the sidereal time $18^h.40^m$ of which the year begins, according to the common method of counting days, when the commencement of the same is reckoned from noon. The culmination of those stars whose right ascensions are less than $18^h.40^m$ does not fall, therefore, on that day denoted by Jan. 0 in the tables, but on the day preceding, and thus to the datum of the day reckoned from noon, must be added a unit, before entering the tables.

The arguments are consequently the following:

1st, if α be $< 18^h.40^m$,

from the beginning of the year to the day on which the right ascension of the sun is equal to α,

$$\text{datum} + k + d + \frac{\alpha - 18^h.40^m}{24^h} + 1;$$

from that time to the end of the year,

$$\text{datum} + k + d + \frac{\alpha - 18^h.40^m}{24^h} + 2.$$

2ndly, when α is $> 18^h.40^m$,

from the beginning of the year to the day on which the right ascension of the sun is equal to α,

$$\text{datum} + k + d + \frac{\alpha - 18^h.40^m}{24^h};$$

and, from that time to the end of the year,

$$\text{datum} + k + d + \frac{\alpha - 18^h.40^m}{24^h} + 1.$$

15. This kind of computation of apparent places is particularly convenient when it is desired to construct an ephemeris of stars' places for a considerable length of time.

If only a single place be required, the following method may be used more conveniently, since thereby the trouble of computing the constants *a*, *b*, *c*, &c. is avoided.

The terms of the precession and nutation are, namely, if the quantity E be neglected:
for right ascension,

$$Am + An \sin\alpha \tan\delta + B \tan\delta \cos\alpha;$$

and, for declination,

$$An \cos\alpha - B \sin\alpha.$$

Put now

$$An = g \cos G,$$
$$B = g \sin G,$$
$$Am = f,$$

then will these terms be, for the right ascension,

$$f + g \sin(G + \alpha) \tan\delta;$$

and, for the declination,

$$g \cos(G + \alpha).$$

Moreover we have for the aberration in right ascension and declination, by No. 14 of Section 2,

$$h \sin(H + \alpha) \sec\delta,$$

and

$$h \cos(H + \alpha) \sin\delta + i \cos\delta,$$

$$(\text{where } h \sin H = C, \quad h \cos H = D, \quad \text{and } i = C \tan\epsilon),$$

so that the complete formulæ for the apparent place of a star, are

$$\alpha' = \alpha + f + g \sin(G + \alpha) \tan\delta + h \sin(H + \alpha) \sec\delta + \tau\mu,$$

$$\delta' = \delta + g \cos(G + \alpha) + h \cos(H + \alpha) \sin\delta + i \cos\delta + \tau\mu'.$$

Note. On the computation of apparent places compare the preface to Bessel's *Tabulæ Regiomontanæ*, pp. 24 and 29, &c.

END OF THE FIRST PART.

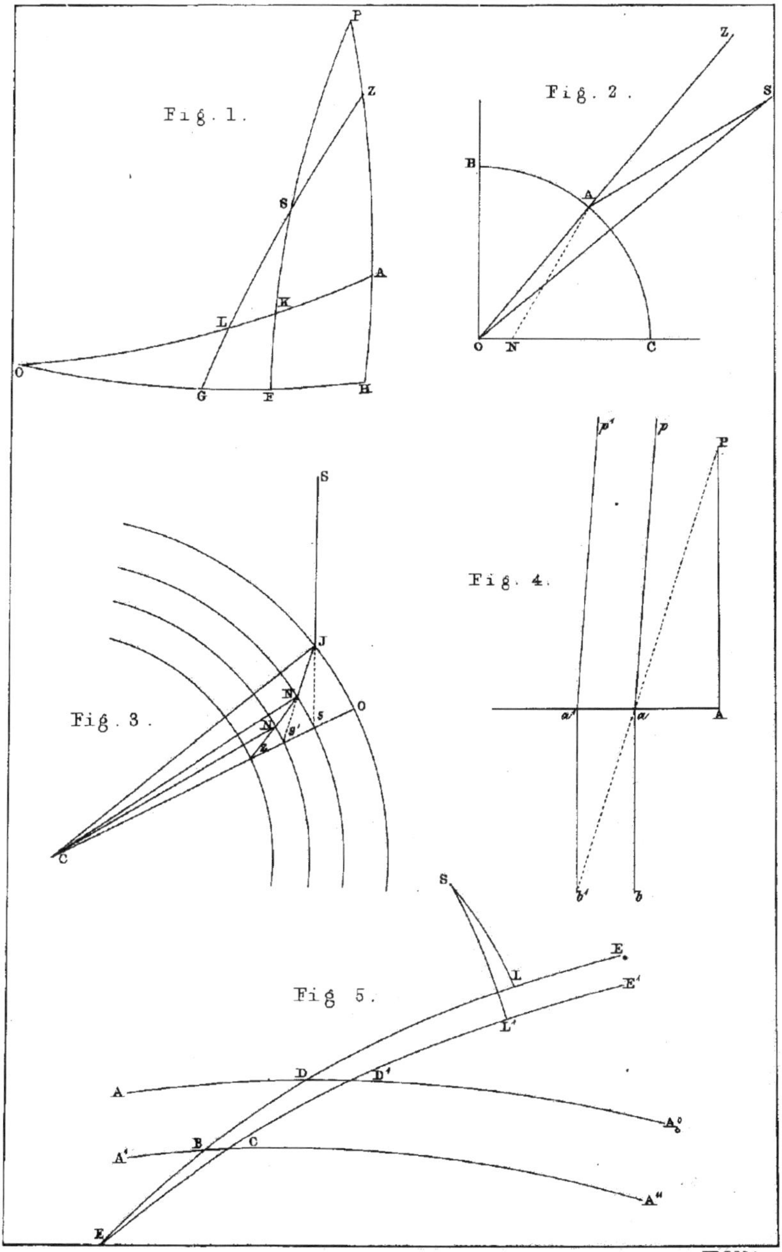
Fig. 1.
P
Z
S
A
K
L
O
G
F
H
Fig. 2.
Z
S
B
A
O
N
C
Fig. 3.
S
J
N
O
N
s
s'
Z
C
Fig. 4.
p'
p
P
a'
a
A
b'
b
Fig 5.
S
E
L
E'
L'
D
D'
A
A°
A'
B
C
A''
E

J.R.Jobbins

Spotted something that isn't quite right?

Let us know and we'll fix it for future reprints. And as a thank you for your diligence, we'll send you a free Sothis book of your choice.

publisher@sothis.press

This page intentionally left blank

Made in the USA
Coppell, TX
05 November 2026

11365025R00129